High Performance Computing on Vector Systems 2010

Michael Resch • Katharina Benkert • Xin Wang •
Martin Galle • Wolfgang Bez • Hiroaki Kobayashi •
Sabine Roller
Editors

High Performance Computing on Vector Systems 2010

Springer

Editors

Michael Resch
Katharina Benkert
Xin Wang
Höchstleistungsrechenzentrum
Stuttgart (HLRS)
Universität Stuttgart
Nobelstraße 19
70569 Stuttgart
Germany
resch@hlrs.de
benkert@hlrs.de
xin.wang@hlrs.de

Martin Galle
Wolfgang Bez
NEC High Performance
Computing Europe GmbH
Prinzenallee 11
40459 Düsseldorf
Germany
mgalle@hpce.nec.com
wbez@hpce.nec.com

Hiroaki Kobayashi
Cyberscience Center
Tohoku University
Aramaki-Aza-Aoba 4F
980-8578 Sendai
Japan
koba@isc.tohoku.ac.jp

Sabine Roller
German Research School
for Simulation Sciences
Schinkelstr. 2a
52062 Aachen
Germany
s.roller@grs-sim.de

Front cover figure: Snapshot of the acoustic pressure field generated by a globally unstable hot jet. The blue spheres indicate virtual microphones. Illustration by Institute of Aerodynamics, RWTH Aachen University, Aachen, Germany

ISBN 978-3-642-44653-5
ISBN 978-3-642-11851-7 (eBook)
DOI 10.1007/978-3-642-11851-7
Springer Heidelberg Dordrecht London New York

Mathematics Subject Classification (2010): 68Wxx, 68W10, 68Mxx, 68U20, 76-XX, 86A10, 70FXX, 92Cxx

Softcover re-print of the Hardcover 1st edition 2010

Cover design: WMX Design, Heidelberg

Printed on acid-free paper

Springer is part of Springer Science+Business Media (www.springer.com)

Preface

This book covers the results of the 11th and 12th Teraflop Workshop and continued a series initiated by NEC and the HLRS in 2004. As part of the Teraflop Workbench, it has become a meeting platform for scientists, application developers, international experts and hardware designers to discuss the current state and future directions of supercomputing with the aim of achieving the highest sustained application performance.

The Teraflop Workbench Project is a collaboration between the High Performance Computing Center Stuttgart (HLRS) and NEC Deutschland GmbH (NEC HPCE) to support users to achieve their research goals using High Performance Computing. The first stage of the Teraflop Workbench project (2004–2008) concentrated on user's applications and their optimization for the 72-node NEC SX-8 installation at HLRS. During this stage, numerous individual codes, developed and maintained by researchers or commercial organizations, have been analyzed and optimized. Several of the codes have shown the ability to outreach the TFlop/s threshold of sustained performance. This created the possibility for new science and a deeper understanding of the underlying physics.

The second stage of the Teraflop Workbench project (2008–2012) focuses on current and future trends of hardware and software developments. We observe a strong tendency to heterogeneous environments at the hardware level. At the same time, applications become increasingly heterogeneous by including multi-physics or multi-scale effects. The goal of the current studies of the Teraflop Workbench is to gain inside into the developments of both components. The overall target is to help scientists to run their application in the most efficient and most convenient way on the hardware best suited for their purposes.

The papers in this book draw a bow from leading edge operating system development to the needs and results of real life applications in various scientific areas. They put the different views of hardware specialists, supercomputing centers, and users on a common topic up to discussions, namely to enable and facilitate leading edge scientific research.

The work in the Teraflop Workbench project gives us insight into the applications and requirements for current and future HPC systems. We observe the emergence of multi-scale and multi-physics applications, the increase in interdisciplinary tasks and the growing tendency to use today's stand-alone application codes as modules in prospective, more complex coupled simulations. At the same time, we notice the current lack of support for those applications. Our goal is to offer an environment that allows users to concentrate on their area of expertise without spending too much time on computer science itself.

We would like to thank all the contributors of this book and the Teraflop Workbench project. We thank especially Prof. Hiroaki Kobayashi for the close collaboration over the past years and are looking forward to intensify our cooperation in the future.

Stuttgart, May 2010

Katharina Benkert
Michael M. Resch

Contents

Part I
Operating System and Software for Large Scale Systems

Light-Weight Kernel with Portals

Erich Focht, Jaka Močnik, Fredrik Unger, Andreas Jeutter, Marko Novak

Abstract With continuously growing numbers of nodes and CPU cores cluster scalability is becoming a more and more significant problem in high performance computing and several approaches are taken to improve it. On the hardware level, operating system level and in the communication model new approaches have been developed. Specialization of cluster nodes, introduction of light-weight kernels and new communication abstraction are all steps to increase the efficiency of compute clusters. Extending the light-weight kernel (LWK), Kitten, with RDMA capable Infiniband network interface support and developing Portals on top of that interface brings improvements to the current compute model. Furthermore, in preparation for running parallel jobs on the light-weight kernel a new Open MPI component was added as an alternative to the currently available OOB/TCP component. This component eliminates the need to have a TCP/IP software stack available on the compute nodes. It is based on the Sandia Portals 3.3 network abstraction and message passing interface.

1 Introduction

Clusters are growing larger with respect to the number of nodes and cores. Looking at the challenges of scaling applications to large clusters, different levels of prob-

Erich Focht, Fredrik Unger
NEC HPC Europe, Hessbrühlstr. 21b, 70565 Stuttgart, Germany
e-mail: efocht@hpce.nec.com, funger@hpce.nec.com

Jaka Močnik, Marko Novak
XLAB Research, Pot za Brdom 100, 1000 Ljubljana, Slovenia
e-mail: jaka.mocnik@xlab.si, marko.novak@xlab.si

Andreas Jeutter
High Performance Computing Center Stuttgart (HLRS), Nobelstr. 19, 70569 Stuttgart, Germany
e-mail: jeutter@hlrs.de

M. Resch et al. (eds.), *High Performance Computing on Vector Systems 2010*,
DOI 10.1007/978-3-642-11851-7_1, © Springer-Verlag Berlin Heidelberg 2010

lems arise. At the hardware level, in order to solve resource problems, cluster nodes become specialized for different tasks (e.g. Fig. 1). A natural step is to offload tasks like I/O, access and management from compute nodes to make them as efficient as possible. A service node does for example not need an as powerful a CPU as a compute node, but might need additional PCI devices to connect to storage devices.

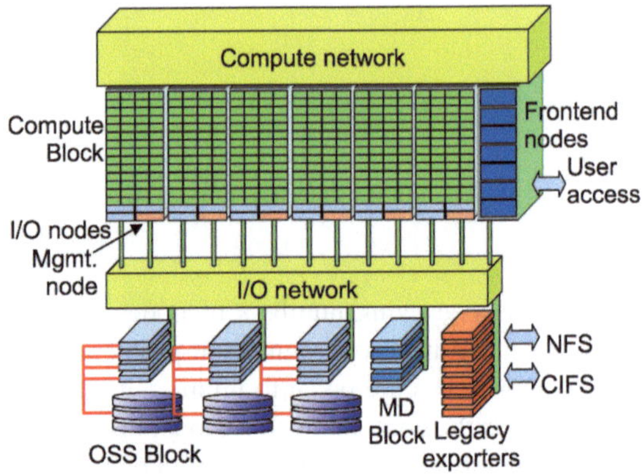

Fig. 1 Specialization of cluster nodes

At the operating system level new approaches are also needed. The use of a full desktop or general-purpose server operating system like a full Linux distribution installation is not the optimal choice for a server whose task is to just run an HPC application with maximum performance. One approach to this is to switch the compute node kernel to a light-weight variant, where the kernel just provides mechanisms to virtualize hardware resources and enforce them, but the policies of enforcement are implemented in user space, as done for example for Catamount [1] or in microkernels [2]. The advantages are various: a simple and small kernel code base is easy to debug and maintain. Moving out complex code to user space (into a single process control thread or various user space services), like scheduling policy code, process management code in general or filesystem implementations (provided as library code), allows for quick modifications and adaptation of the code without touching and modifying the kernel. This has the potential to speed up innovation while keeping a solid and stable basis. The most important advantage is the customization of the system software to the node's task, and the focus clearly is in reducing the OS noise in order to increase the scalability of massively parallel HPC applications.

In last year's report on the progress of the SX-Linux project [6] we described the efforts done to port the Kitten LWK [3] to the SX vector architecture. In the mean time we moved our LWK related efforts towards the x86_64 architecture and aim at providing components that help improving the scalability of clusters built with off the shelve motherboards, CPUs and Infiniband interconnect.

At the communication level simplification of the layers and a replacement of the socket model is needed. Support of the Infiniband communication hardware with abstractions that match its RDMA capabilities is important. In this respect Portals from Sandia National Labs provides a suitable model and an optimal replacement of TCP/IP for system software communication needs.

For MPI support the initial target implementation was OpenMPI, and enablement of InfiniBand for all the communication layers. One step in the development was to remove OpenMPI's dependence on the Berkeley sockets for the internal out-of-band (OOB) communication and replace it by use of Portals. The application level also needs I/O support and as the new model implies specialized I/O nodes an I/O forwarding layer is needed. The latter is the subject of another paper [4].

Section 2 presents the Kitten light-weight kernel, and describes the effort required to port linux-based OFED Infiniband support to the Kitten kernel. Preliminary benchmarks of the ported Infiniband communication support are given. An introduction to the Portals communication API is provided in Sect. 3, and the optimizations of the core library and implementation of a high-performance Infiniband Portals driver (NAL) are presented, followed by benchmarks of Portals communication over Infiniband. Section 4 describes an adaptation of the Open MPI library that allows to run MPI without the need for TCP/IP-based out-of-band communication. The paper concludes with a short summary of current status and the directions for future work.

2 Kitten

A typical cluster node of today runs a full Linux kernel, normally from vendors like Redhat or SUSE. This kernel is normally configured to be able to work with a large variety of hardware and provides a lot of system services support. While these system services might be important in some case, they are far more commonly a problem for an HPC application as they take valuable CPU cycles from the application and creates an imbalance between nodes at synchronization points of the application.

A light-weight kernel provides the minimal functionality needed for a compute application to run: hardware recognition, physical and virtual memory management, task management, interrupts and system calls. The design choice behind the light-weight kernel reduces OS overhead in favor of the application. Some functionality like file system support or special scheduling policies can be implemented outside the kernel in user space threads or libraries. Remaining compatible with the Linux kernel is also important in order to be able to run ISV applications and limiting the work to port an application to the light-weight kernel.

The light-weight kernel Kitten was created at Sandia National Labs and is developed as an open source project [5]. It uses simple task scheduling and a deterministic memory mapping (virtual to physical). It has a limited number of Linux ABI system calls and some light-weight kernel specific system calls. The development target is to keep the light-weight kernel as simple and powerful as possible.

2.1 OFED

The OpenFabrics Alliance (OFA) develops the OpenFabrics Enterprise Distribution (OFED) which is an open-source software stack providing support for new RDMA network interfaces, like Infiniband. The Kitten light-weight kernel networking initially relied on the light-weight IP stack (LWiP) and an older port of OFED 1.3. This is not enough to provide HPC applications access to the newest Infiniband network interfaces. OFED 1.4.1 was initially added to the kitten source tree, and eventually also 1.4.2 and 1.5.1. To provide support for the OFED stack in Kitten, the kernel had to be extended in several ways.

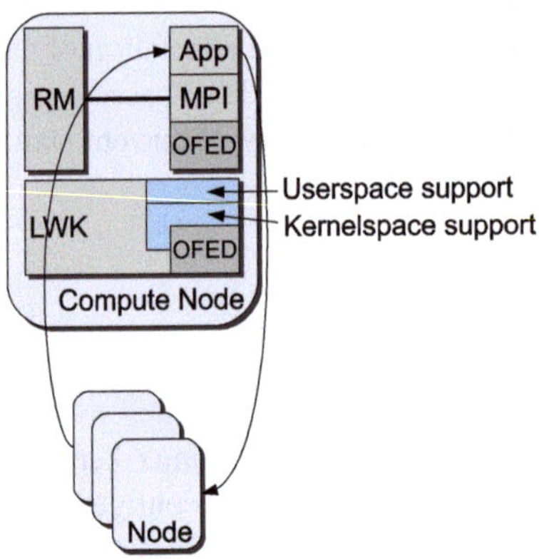

Fig. 2 OFED development in Kitten

The first steps that had to be taken once the OFED stack and drivers were added was to add the kernel support needed by the drivers to be able to communicate with the card, and the outside world. A small Linux API emulation layer provided this infrastructure to enumerate the PCI bus. Driver registration code and DMA support were added. The kernel driver is divided into several parts. Core that provides the overall Infiniband support, mlx4 and mthca provide drivers for two Mellanox cards: Mellanox ConnectX, and Mellanox Infinihost, respectively. Other drivers from Linux source tree can be ported with a minimum of effort. After the driver core components were enabled together with the specific card driver the card could communicate with the Infiniband fabric.

The next step was to provide the userspace libraries with the view of the card they are expecting from Linux by adding the relevant device files and sysfs entries. A simple sysfs for registration and a simple devfs for initial setup of the drivers were added. System calls poll and mmap were added to provide the user libraries the means of communication with the card that is used in Linux. Thanks to these small additions an Infiniband application, statically compiled under Linux can run without any modification on a Kitten cluster node, as all the interfaces to the kernel are provided to userspace in the same way as on a fully-fledged Linux kernel.

2.2 Benchmarks

Figures 3 and 4 show the latency and bandwidth of different Infiniband operations (send, RDMA read and RDMA write) as a function of user payload size on Kitten and Linux. Linux performance is slightly better, which is attributed to the overhead and suboptimal implementation of Linux API wrappers around core Kitten kernel functionality.

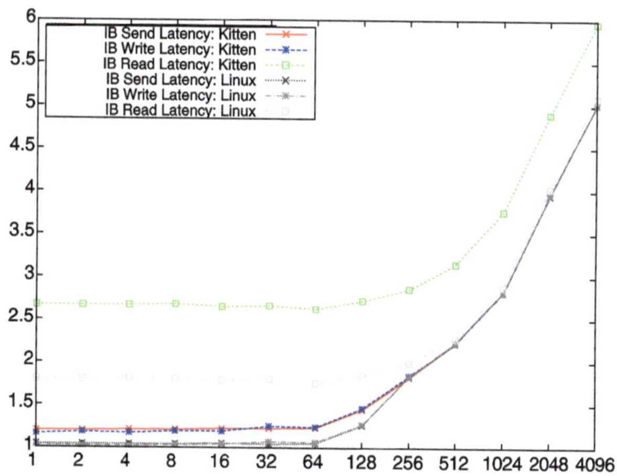

Fig. 3 IB Latency [μs]

Note that the bandwidth measurements fail to reach the expected values due to the benchmark implementation. Bandwidth achieved by Portals (see Fig. 6) is thus higher in spite of the additional overhead.

3 Portals

In order to substitute for socket-based IP communication, which the light-weight kernel does not support, a suitable network abstraction to be used for system software and optionally for parallel communication as well had to be provided.

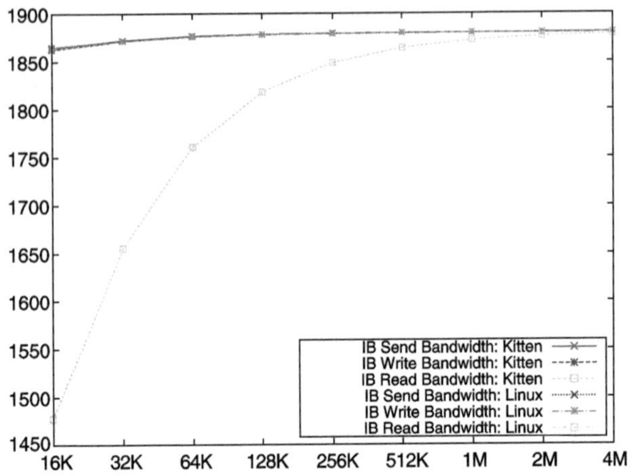

Fig. 4 IB Bandwidth [MB/s]

In selecting such an abstraction, the following criteria were considered:

- fully asynchronous design,
- straightforward mapping to native APIs of current communication hardware,
- simple API for the developer,
- existing support in common HPC software,
- open-sourced codebase to leverage in our work.

3.1 Portals: A Brief Introduction

Portals [7] API provides an abstraction of RDMA communication. Designed by Sandia National Laboratories, Portals are commonly used on Cray machines, and a number of HPC software components supports Portals as the communication abstraction (OpenMPI [12], PVFS [11] via BMI, GASnet [10], etc.).

Portals API is based on simple *get and* put operations on registered memory regions of communicating processes: get fetches data from a remote process' memory into local memory while put transfers local memory contents into a remote process' memory. Every Portals operation (in addition to local memory area to use as source or destination) uniquely references a remote memory area on the network involved in the operation with:

- a global process ID (consisting of remote node ID, NID, and node-local process ID, PID),
- the portal index (every process can use multiple portals, and associate different memory regions with each),

- 64 bit sequence of *match bits* used to identify the memory region within a single portal.

Portals provide an event-driven model for the user application, which invokes an operation and then waits for events that denote progress and finally (un)successful termination of an operation.

A typical sequence of operations involved in a Portals data exchange between two (an initiator and a target) processes thus consists of:

- both processes initializing the network interface used for communication,
- target process opening the portal, registering one or more memory regions, attaching them to the portal *match list*, and associating each memory region with a sequence of *match bits* which are later used by remote processes to identify the memory region to operate on,
- initiator node invoking an operation on a remote memory area identified by match bits value consistent with the one used by the target process,
- initiator node polling for event that will denote that the transfer into remote memory area has finished successfully,
- target node polling for events that denote that the transfer into its local memory area has finished successfully.

While Portals 4.0 specification [8] has been available for quite a while now, the only freely available implementation is the reference implementation of Portals 3.3 specification [9], which was used as the base for our work.

3.2 Optimizing Portals for LWK

The reference Portals implementation consisted of three separate components:

- the API, providing user interface to the Portals functionality,
- the library, implementing the Portals communication semantics,
- the NAL (*Network Abstraction Layer), performing the actual communication over an arbitrary type of interconnect.*

Any of these components can run in a separate address space (user space, kernel space, and, in case of smart NICs, such as the Cray Seastar interconnect, even NIC address space) and the reference implementation used message passing paradigm for communication between API and library components.

As the implementation presented in this paper is aiming at an all-user-space Portals in accordance with the principle of keeping the kernel small and simple, the message passing approach to communication between API and library was removed in order to minimize overhead, and all kernel-related parts of the code were removed.

3.3 A High-Performance Infiniband NAL

The only NAL implemented in the original sources used TCP/IP communication. Therefore, a new NAL for Infiniband interconnects, common in modern HPC systems (207 systems in the Top 500 list, 41.4%, as of June 2010), was implemented.

Infiniband LID (Local ID) is used as Portals NID, and OS process ID is used as Portals PID. A (LID, PID) pair uniquely identifies a process in the cluster. Infiniband Reliable Connected (RC) Queue Pairs (QP) are used for communication between process. One RC QP is used for each remote process being communicated with.

Connections between processes are established using Infiniband Connection Management (IB CM) protocol. Each process establishes a listening CM ID on startup (the node-local process ID is used as the CM ID). A connecting process can use LID and remote PID to address a remote process via CM protocol and exchange connection parameters. During a successful CM handshake, QPs for the connection are created by each of the two processes involved, and QPs are connected. Such a connection is established at the first communication attempt between a pair of nodes and is preserved and reused for all further communication between the said peers until explicitly closed.

At the time of connection establishment, send and receive buffers are allocated and receive buffers posted. The send and receive buffer pools are dynamically increased during lifetime of the connection if the communication pattern requires an increase. With a default set-up, each connection requires approximately 0.5 MB of communication buffer memory—the exact amount may be tuned at process start by setting appropriate environment variables that determine individual buffer sizes and their initial numbers. Total memory consumption is dominated by buffers and increases linearly with number of established connections. This memory also proves the main bottleneck for scaling the number of communicating processes: communicating with 8192 processes (all-to-all communication in a 1024-node cluster with 8 cores per node) would require $8 \times 8191 \times 0.5$ MB $= 32$ GB of buffer memory per node.

After establishing the connection, Infiniband send, RDMA read and RDMA write operations are used to transfer Portals messages, consisting of Portals header and user payload.

The NAL packs fixed-size Portals header which amounts to 80 bytes in the original implementation to a variable sized header, with an average size of 24 bytes. 32 bits of header that is common for all operations is transported as an *immediate value* in the Infiniband header, while the rest is transported as Infiniband packet payload.

The header is always transported with the Infiniband send operation into a preposted receive buffer of the target process.

In case of small messages, the user payload is also transferred with the send operation and copied into the user buffers on the target node (i.e. *eager transfer*). In case of large messages, the header is followed by information required for Infiniband RDMA transfers (remote address and remote key) instead, allowing the target node to perform a RDMA operation directly into the user buffer, avoiding memory copying at the price of an additional Infiniband operation (i.e. *rendezvous transfer*).

3.4 Benchmarks

Figures 5 and 6 present latency and bandwidth of Portals with the Infiniband NAL between two Kitten nodes as a function of user payload size. Linux performance is also presented for comparison.

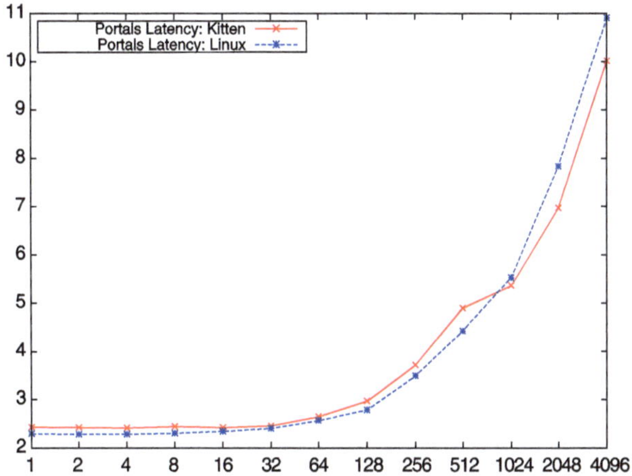

Fig. 5 Portals Latency [μs]

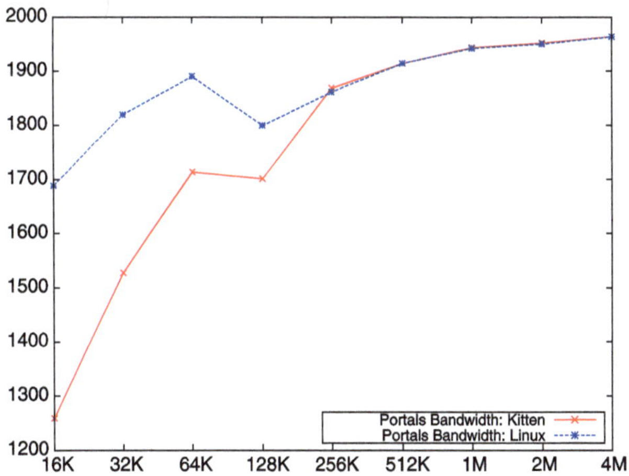

Fig. 6 Portals Bandwidth [MB/s]

The Portals latency results are very close to the native Infiniband latency values. For small messages Kitten shows slightly larger latency values than Linux, a fact attributed to the overhead of the unoptimized Linux compatibility layer that was needed for adding Infiniband drivers support (see Fig. 3 as well). However, messages above 512 bytes required fragmentation with the buffer sizes used in benchmark setup (i.e. a single portals payload is sent with multiple Infiniband send operations): there, Kitten clearly outperforms Linux due to more CPU time being available for the user-space task. All the latency results shown here use eager transfer.

The Portals bandwidth results are obtained using rendezvous transfer mode. In case of large messages (256 K and up) Kitten and Linux reach the same bandwidth. With lower bandwidth, Kitten performed significantly worse. The reason for this is being investigated, but is unclear at the moment. As Infiniband RDMA read operation used in this kind of transfer performs equally well on Linux and Kitten (refer to Fig. 4), the cause must be in the Portals user-space code.

4 MPI

All parallel applications, regardless of the MPI flavor used, have massive communication requirements. The Open Runtime Environment (ORTE) as part of Open MPI comprises infrastructure programs and libraries to start and run parallel applications on many nodes simultaneously. The most important components involved in running an Open MPI application are mpirun, the compute node daemon orted and the MPI library.

The communication interface to be used for the application's MPI messages, can be selected on the command line, whereas the ORTE internal infrastructure communication is solely based on TCP/IP.

4.1 ORTE Job Preparation and Startup

Preparing a parallel application for running on many nodes includes starting the job by invoking mpirun. This is either done by the user within an interactive session or through a batch system, like PBS/TORQUE. Mpirun expects the hostnames of the compute nodes (CN) dedicated to the job. Usually they are passed to mpirun on the command line or within a host file.

Mpirun starts the job by spawning ssh processes that connect to ssh daemons (sshd) on the specified nodes. Once the ssh path is established, sshd forks a new process and executes the ORTE daemon (orted). Mpirun waits until orted becomes active. The ORTE daemon activates its Remote Messaging Layer (RML) component, which checks for available OOB (Out Of Band) components. All programs and tools participating in the RML communication must be set up to use the same OOB component. Currently, only the OOB/TCP component can be selected since it is the only one available. This component uses Berkeley sockets on TCP/IP to send and receive messages.

When mpirun starts the daemon on the compute nodes it takes its own hostname and the port where the local OOB/TCP socket is listening and puts them on the dae-

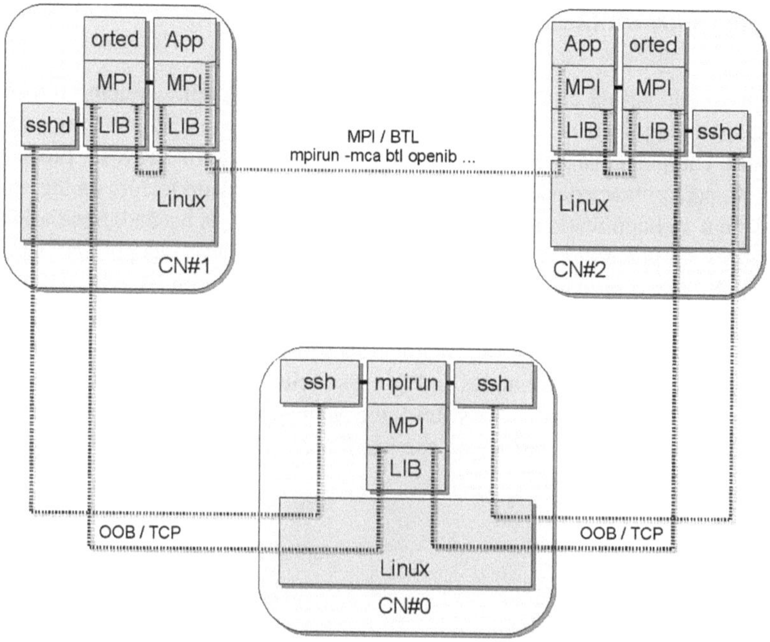

Fig. 7 ORTE OOB/TCP: common startup scenario

mon's command line. The daemon loads the local OOB/TCP component, opens and initializes it and passes the hostname/port parameters to it. The component uses this information and initiates a TCP connection to mpirun. Mpirun and the daemons utilize this message path to exchange job information like job size (retrieved from the MPI application with a call to MPI_Comm_size()), node rank (MPI_Comm_rank()) and the name of the MPI application.

In the next step the daemon starts the specified MPI application and puts its own hostname and the port where it is listening for inbound connections into the environment of the MPI application. The MPI application typically contains a call to MPI_Init() at the very beginning of the program. This call leads to the activation of the underlying ORTE infrastructure. Equivalent to the daemon, the MPI application loads the OOB component and initializes it. The OOB/TCP component reads the environment, initiates a connection to the daemon and sends a message that it is ready. The daemon forwards this message to mpirun. Mpirun registers the successful launch of the application and sends job information to the application (Fig. 7).

In the further process, the daemon relays information between the MPI application and mpirun and also forwards output that the application writes to standard out through OOB/TCP. Mpirun keeps a dictionary of all involved nodes and processes, their rank and status. The MPI application can request those data and use them to exchange messages directly with other processes without having to use the RML component anymore.

4.2 Job Start on the Light-Weight Kernel

A light-weight kernel differs from a general purpose operating system, in that it is dedicated and optimized to be used for running parallel applications. In our special case, the computer hardware uses Infiniband NICs that utilize RDMA procedures for high–performance data exchange. To reflect this hardware feature on the operating system and application level a suitable software suite is needed. Sandia Portals provides such procedures, which allow direct memory access via put and get commands. As stated previously the ORTE OOB relies solely on TCP/IP, thus a new OOB/Portals component, equivalent to the OOB/TCP component, had to be developed. In the scenario described above, the SSH software suit, which is used to launch the daemon on the compute nodes and that also relies on TCP/IP, must be replaced. This new software is described as Resource Manager (RM) on the compute nodes and Resource Controller (RCtl) on the head node (HNP, CN#0) where mpirun is running.

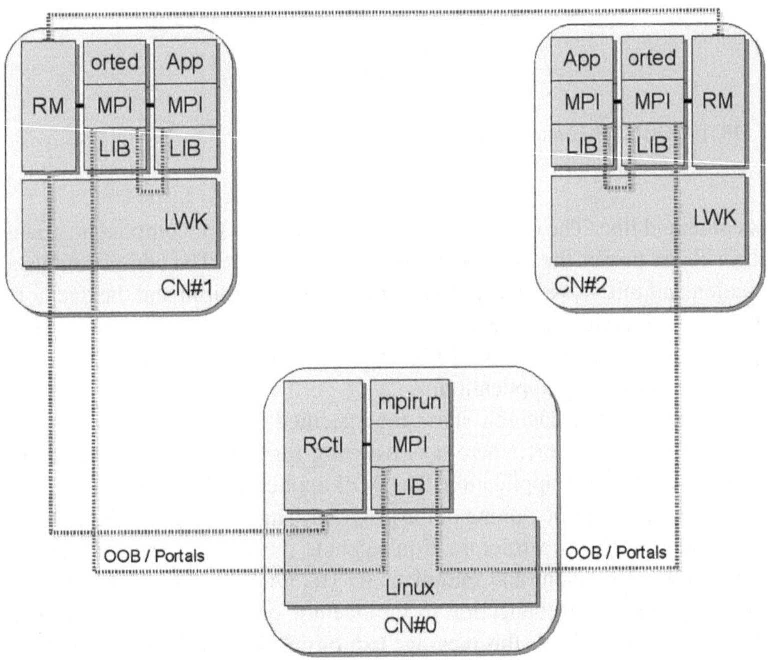

Fig. 8 OOB/Portals on the LWK: start via RM

In the light-weight kernel (LWK) scenario, all compute nodes boot, activate and run RM. When compared to a conventional SysV startup sequence this amounts to replacing the init process with the resource manager process.

On the initial node (CN#0) mpirun forks a new process and runs the resource controller. All resource managers are connected in a tree-like structure. Once all compute nodes are booted and active, the resource controller sends the orte daemon executable via network to all directly connected resource managers. Upon receiving the executable each resource manager stores it in a temporary file system, prepares a command line and forks a new process to run the executable. The daemon retrieves the command line parameters and activates the new OOB/Portals component, which uses the contact information and establishes a connection to mpirun via Portals (Fig. 8).

From this point on, the scenario behaves like previously described. The compute nodes receive job information, launch the MPI application and forward standard output stream via OOB/Portals.

4.3 Architecture of OOB/Portals

An application independent remote procedure call layer (RPC) has been put on top of Portals to hide the handling of memory descriptors, match entries and other Portals API internals.

The RPC layer features a trivial API that allows asynchronous sends and receives. To receive data, a register function is set up, which installs a receive buffer and a callback handler. This allows receives to run in the background and trigger callback functions when data is ready to be processed. Sends are also asynchronously executed in the background. After a successful send, the buffer is freed automatically.

RPC uses two different methods to send data to a peer. A size threshold determines which method is to be used. Messages below the threshold are sent directly via a Portals put command. Messages above the threshold are sent by first sending a notification message to the peer and the peer in turn fetches the message from the sender via the Portals get function. This schema allows messages of arbitrary size to be sent, while maintaining a fair compromise between memory footprint and performance.

5 Conclusion and Future Work

The paper describes the effort done on the way to developing a more scalable system environment for clusters that have become commodity in HPC: built on x86_64 architecture with Infiniband interconnect. While proprietary hardware (e.g. from CRAY or IBM) already has support for some type of light-weight operating system, this is usually developed as closed source and unavailable for off-the-shelve built systems. Our developments aim at improving this situation and expanding the toolset available for improving scalability on high end HPC systems.

The first development described was focused on adding support for Infiniband devices to the Kitten light-weight kernel. This effort will continue with the integration of further hardware and optimization of the LWK linux compatibility infrastructure.

For the user-level system software we worked on a network abstraction and wrote an Infiniband network abstraction layer for the Portals networking stack. It enables writing cluster system software without TCP/IP and sockets, that are intentionally missing in the Kitten LWK. As a next step we plan to improve the scalability of the Infiniband NAL with regard to memory consumption: a hybrid UD-RC approach [13] as used in MVAPICH is a good candidate.

The first steps towards user applications were taken by adapting Open MPI to the missing TCP/IP stack on the Kitten LWK and developing a first working implementation of OOB/Portals. Future goals include fully integrating the component into the Open MPI project as well as improving the application startup scalability and adding support for starting MPI applications inside the Kitten LWK.

References

1. S. Kelly and R. Brightwell: Software architecture of the lightweight kernel, Catamount. In 2005 Cray Users Group, May Annual Technical Conference. 2005.
2. J. Liedtke: Toward real microkernels. Published in Communication of the ACM (CACM), vol 39/9, pp. 70–77, September 1996.
3. J. Lange, K. Pedretti, T. Hudson, P. Dinda, Z. Cui, L. Xia, P. Bridges, A. Gocke, S. Jaconette, M. Levenhagen, and R. Brightwell: Palacios and Kitten: New High Performance Operating Systems for Scalable Virtualized and Native Supercomputing. In IPDPS '10: Proceedings of the 24th IEEE International Parallel and Distributed Processing Symposium (19–23 April 2010).
4. Erich Focht, Thomas Großmann, and Danny Sternkopf: I/O Forwarding on NEC SX-9. High Performance Computing on Vector Systems 2010.
5. Kitten light weight kernel project page, http://code.google.com/p/kitten.
6. E. Focht, J. Močnik, F. Unger, D. Sternkopf, M. Novak, and T. Grossmann: The SX-Linux Project: A Progress Report. High Performance Computing on Vector Systems 2009, pp. 79–96. DOI: 10.1007/978-3-642-03913-3_8.
7. Sandia Portals, http://www.cs.sandia.gov/Portals/.
8. Rolf Riesen, Ron Brightwell, Kevin Pedretti, Brian Barrett, Keith Underwood, Arthur B. Maccabe, and Trammell Hudson: The Portals 4.0 Message Passing Interface. Technical Report, Sandia National Laboratories, April 2008.
9. Rolf Riesen, Ron Brightwell, Kevin Pedretti, Arthur B. Maccabe, and Trammell Hudson: The Portals 3.3 Message Passing Interface. Technical Report, Sandia National Laboratories, April 2006.
10. D. Bonachea, P. Hargrove, M. Welcome, and K. Yelick: Porting GASNet to Portals: Partitioned Global Address Space (PGAS) Language Support for the Cray XT, CUG 2009.
11. Parallel Virtual File System, Version 2, http://www.pvfs.org/.
12. Open MPI, http://www.open-mpi.org/.
13. M. Koop, T. Jones, and D. K. Panda: MVAPICH-Aptus: Scalable High-Performance Multi-Transport MPI over InfiniBand. IEEE International Parallel and Distributed Processing Symposium (IPDPS '08), Miami, Florida, April 2008.

Towards an Architecture for Management of Very Large Computing Systems

Jochen Buchholz, Eugen Volk

Abstract Managing very large computing systems with up to 100.000 nodes has become a very complex issue. Existing tools reach their limits especially for High Performance Computing (HPC) resources because they are slightly different from other compute resources. First we will introduce the specific HPC obstacles and what we suppose to be challenges for future resources to support the system management. After that we propose the framework designed in scope of the TIMaCS Project (http://www.timacs.de). Assuming that we once have a corresponding solution implemented we will show how this solution can change administration far beyond the current situation. This is separated into a more technical part describing how the administration can be simplified or where we can add new capabilities in resources provisioning and a business part where we outline the need for business policy based management and scheduling, and show a possible approach investigating these relationships. In the end we will show what might be possible far beyond the scope of the project.

1 Introduction

HPC has evolved in the last years from an appearance on the fringes where it was used only for some specific simulations like weather forecast to a heavily used tools which are fully integrated in the development process of many goods and even used for just in time applications in the medical area. Although the involvement of HPC is often not obvious for the consumer or user, lots of products could not be built or services not offered without HPC, at least at a significant lower level. For example flow simulations for new products—from cars to turbine simulations for hydro power plant—increase product efficiency and reduce their resource (i.e. fuel) con-

Jochen Buchholz, Eugen Volk
High Performance Computing Center Stuttgart (HLRS), Nobelstr. 19, 70569 Stuttgart, Germany
e-mail: buchholz@hlrs.de, volk@hlrs.de

M. Resch et al. (eds.), *High Performance Computing on Vector Systems 2010*,
DOI 10.1007/978-3-642-11851-7_2, © Springer-Verlag Berlin Heidelberg 2010

sumption. Instead or traditional physical experiments over a long time with lots of different examples they are simply simulated with an even broader variety of starting parameters.

1.1 Specific Challenges in HPC

HPC providers are currently challenged by several general changes at the same time which are not covered by existing administration frameworks. They can be divided into three areas, the technical part including all actions to run a resource, the usage of these resources and functional requirements.

The technical part contains everything from hardware setup over software installation processes to all necessary actions needed to run the resource itself. The main changes within are hardware developments towards many core systems and heterogeneous clusters which will soon replace most homogeneous cluster. This will include number and type of processing units (scalar, vector, graphic), memory size, network interconnect and so on. This will increase the complexity somehow but the main problem is scalability. To keep simple usage models the administration tools need to cope with these changes in addition with the increasing number of nodes within each HPC resource.

The second part is even more in flux since more and more user groups—from science, industry, health etc.—have various and higher requirements on HPC resource usage: they want to use HPC resources to advance their work by speed up, obtaining a more detailed view, shortening response time etc. Higher and various requirements of users' and user groups on HPC resources are accompanied by an increasing number of used software products, resulting in higher complexity to fulfill all needs. Here the main focus will be the increased complexity which has to be managed to provide a platform for all user groups.

In addition there are also new functional requirements like external data storage for extremely large data sets or urgent computing for medical purposes which are not covered by traditional HPC resources. They also increase complexity but the main challenge is in many cases that existing procedures and systems can't handle these requirements. So traditional HPC provisioning uses scheduling to reach a very high resource utilization or job-throughput. But for urgent computing you need either idle resources or you have to stop already running jobs. So you will raise some conflicts in doing this, since the owner of the stopped job needs to submit it again with additional waiting time. Last but not least the usage of HPC resources is changing to be much more dynamic. In addition to changing users' requirements on HPC resources, the spectrum of jobs and their complexity is changing as well: users change their providers more often, new applications have to be supported after a short lead-in time, the mixture of submitted jobs changes very often; this should be reflected as well.

In order to provide HPC resources with sufficient quality under the given environmental conditions and business constrains, it is obvious that the system administration has to be technically supported to be able to meet the challenges. For this the current system status and deviations must be detected and visualized. For systems with a low number of nodes it is possible to use simple status lists to detect errors manually but for future systems with 100.000 of nodes or cores with independent operating systems this detection needs to be automatized as well otherwise additional staff members are required. If it is possible to integrate management capabilities for higher complexity in technical solutions, then the administration group will have more time to optimize the systems, foresee new challenges and be prepared even for spontaneous user requirements. Without additional technical support, administrators will soon—and many are already at this state—spend most of their time only in error handling and adjusting the system. This will lead to the awful situation where they are not able to react on changing user requirements, new usage models, etc. and will result in a decreasing attractiveness for users over time. In general the difficulty is to do much more work with identical human resources. As you can see in the left picture of Fig. 1 the red triangle stands for a default situation where you have a stable relation between the work represented by number of users, nodes and functional requirements represented by the area withing the triangle and the human resources. Without increasing your personnel the size of the triangle is static and in any case you enlarge in one of the dimensions you need do shorten another one. The green triangle represents a massive decrease in supported functional requirements for an increasing number of users an nodes.

So it is obvious that we either have to increase human resources or enable the employees to do more work in the same time. This is what we want to achieve as shown in this paper and done in the TIMaCS Project. The result can be visualized as in the right picture where the size of the triangle has increased and therefore the resource can be enlarge to all three dimensions at the same time. As if that had not been enough with several increasing factors like number of nodes, complexity of applications, number of users etc. the visualization of the current system status will be difficult since there is too much information for simple lists which are used currently. It seems possible to use lists for some hundred rows but there are already systems in production where the limitation to deviations instead of status information will reach few thousand entries. So it is necessary to somehow aggregate these

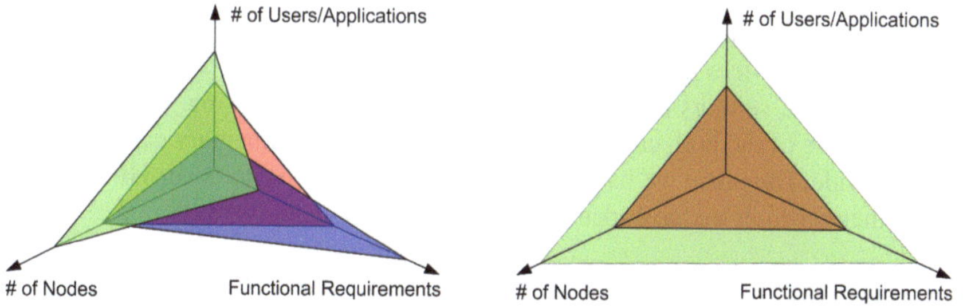

Fig. 1 Human resources and the system's complexity

information by grouping similar deviations or trying to find dependencies and highlight only the independent deviations. The administrator then can drill down on request to find all the dependencies but he is somehow guided through the information overload.

2 Challenges

Between HPC resources and other computing systems we can find some differences regarding hardware (cpu power and network interconnection) as well as in the proposed usage. Since HPC resources are mainly used for (massively) parallel application for simulation the network interconnect is the main difference. For parallel application a very high bandwidth respectively very low guaranteed latency is important to improve the overall efficiency. Other differences like huge memory, small disc or disc less systems have only small effects in comparison to the parallelism. The usage differs from other compute resources in the way that blocks of nodes are used for each user request instead of portions of an node i.e. for a service request where lots of requests from different users are handled by one node at the same time. HPC resources often are used in the node exclusive paradigm, where only one user has access to a node at any time. So the effort for content switching in the operating system can be reduced and the user has full control over the efficiency of his application and don't have to cover side effects from other users. Node sharing could in this way result in the same jitter (see below) problems as any interaction which is done in a non-synchronized way on nodes used by the job.

For gaining information about the current system status, most monitoring and management systems can't be used without modifications for HPC or without losing performance. We don't differ between monitoring and management tools or systems since they are both needed for future systems and therefore had to be combined, but we differ between monitoring and management capabilities instead.

Monitoring includes anything to gain information about the system including aggregation, harmonization of sensor data from different sources and threshold control. It can be seen as the information flow from the bottom level up to the administrator.

Management implies acting on a certain event which can be sent by a monitoring system, by an administrator, or might be time-triggered. The management facility forces services and nodes to be configured according to predefined settings or change these settings depending on the given events. Existing solutions focus mainly on one of the both capabilities and include only rudimentary capabilities on the other side. But since we have to combine both they have to cover HPC specific requirements which we have identified in the following sub-sections.

2.1 Jitter

Current tools for system administration are developed to perform in heterogeneous environments where nodes are not used in a synchronous way. In this scenario for scalability especially monitoring tools try to create sequences for fetching information from sensors so that the events are evenly distributed over time. With this the server running the administration framework has a nearly constant load. In HPC environments the parallelism of applications leads to some undesired side effects. Most application used on HPC resources work a period of time in parallel with using only a portion of data on each node then exchanging some data between the nodes and starting over again. Since any external interaction with a node causes at least a very short delay of the calculation done on this node, the following message exchange leads to delay as well. This results in cascaded delay, as the next calculation step will start later and so on, hence the whole job will be slowed down. This effect will increase with the number of nodes participated in job calculation and number of sensor-data fetched per time period. A detailed explanation is given in Fig. 2.

In scenario A for any type of service provider with serial jobs you can see that the efficiency drops by the time spent for monitoring. In the second scenario B with slightly parallel simulations the job is delayed by the monitoring time multiplied with the number of used nodes for each monitoring interval. Scenario C stands for a massively parallel application with much higher idle time. The exact idle time depends on the rate between simulation, communication and monitoring. Normally simulation time varies a bit for each node since i.e. the used area are not equal or the calculations are more or less complex. So if the monitoring node interferes with a node finishing simulation too early, the monitoring is done instead if the node is idle. Since optimization on application level tries to create simulation block with similar complexity the synchronization will be delayed in the majority of occurrences.

For management actions this negative effect might be insignificant since these actions are done very seldom and in case of errors the running simulations might

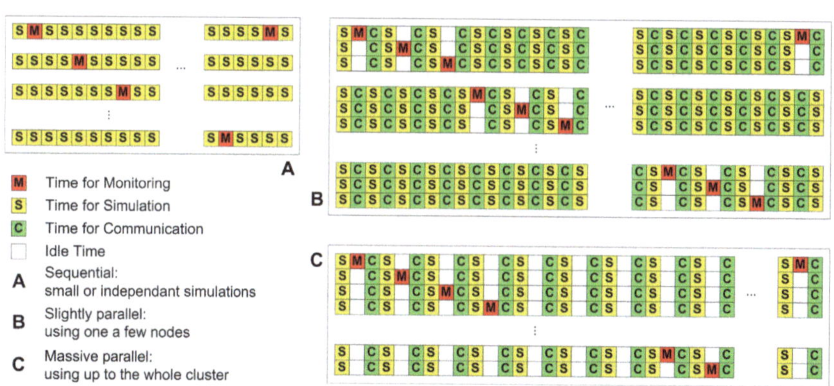

Fig. 2 Example jitter scenarios

have been crashed already so that there are only few negative implications. In case of monitoring actions it is much more difficult since the monitoring is done continuously and on all nodes. Since gathering sensor information is done sequentially, it will be like scenario C where the performance loss is no longer negligible. From the application point of view it is seen as jitter on operation system level. This effect is increasing with the number of nodes. Furthermore, it depends on the circumstances how intense monitoring requests are within the monitoring time period, which occurs every five minutes up to one hour. Shorter and longer intervals between two measurements are mainly used for specific sensors i.e. used to meter current data-throughput on file systems, single points of failure or slowly changing sensors like error counters on hardware level or checking configuration settings. Depending on the time spent for monitoring in comparison to simulation and communication, the idle time can be the major part.

The jitter problem increases linear with the number of observed entities, which are currently nodes or services. In future, these entities might be cores, with operating system running on each one; the negative effect will be based on the number of cores then. This will be a major issue to be covered by future administration frameworks for HPC. A possible approach to solve this issue might be i.e. delaying all monitoring interactions until the simulation has been finished, or synchronizing monitoring at least for a subset of nodes involved in a job. In either cases the monitoring must be aware of the scheduling.

2.2 Scalability

Another issue for a proposed framework is more general for HPC—the scalability. Existing mainstream tools are mainly designed, tested and used in environments with up to a few thousand nodes, which is in comparison to current HPC resources with up to several 100,000 nodes [2] at the top end, and can be reached only if you minimize the number of sensors per node. But can they also withstand the challenge with the intense monitoring on the current and future largest resources with many core systems and even with the possibility to run microkernels on each core [3] Probably monitoring tools need to be somehow adjusted to manage this scaling issue since execution time is not the main problem for monitoring or management actions, but the mass of information which has to be available at one single point because of dependencies between node change with the node-to-job assignment. So the framework has to be able to detect parallel errors caused by one single job. The execution of the action itself can be initiated on several independent machines but the results have to be collected.

One central storage, physical or only logical, allow to analyze the whole system in other ways, enabling identification of trends in usage or error probability, seeking for bottlenecks, searching similar occurrences over time or application type etc. If the information is not linked to each other you cannot find error chains between these parts. Future monitoring tools should be able to fetch data from different partitions if needed should be able at least to show all errors in one view.

2.3 Data Correlation

One trend is towards setting up HPC resources based on mainstream hardware instead of specific HPC hardware. Due to the fact that mainstream hardware has less redundancies or hot-swap capabilities, except discs and power system, is the system more failure-prone. For the whole resources the error rates are increasing along with the increasing number of nodes. For administration you normally need to know the current situation and deviations from the reference values to adjust a single anomaly or correct an error. As a result of the highly increasing number of potential error sources we have to find out when multiple errors depend on each other or at least show possible similarities. This seems to be clear if you have i.e. thousand nodes with the same error and you restart these nodes over and over again because you don't know that they depend on a currently unavailable network file system. In case you can detect these dependencies or somehow describe them the administration framework can hide all dependent errors and therefore focus on the origin of an error. It is not possible to avoid all errors but especially in multi error cases it could be very helpful to filter out such errors, which are in deed only subsequent errors. For this reason, it is also necessary that all information pieces are accessible from one location. With this in mind, it is also possible to create any sort of combined information like aggregation or abstraction of sensor data.

2.4 Error Handling

Current monitoring tools [4] like Nagios [5], Big Brother [6], Zenoss [7] and others are designed to monitor arbitrary systems states and work well within this area. But they offer only very limited management capabilities like reacting on errors or creating abstract sensor information other than simple min/max values for a group of data. Normally they are only able to execute a single script without any control on it. And the script itself can't access the information except accessing e.g. the same database directly.

Capabilities of current management tools are also limited; they are able to force a system to reach certain reference state. For setup and system changes this might be sufficient, but it is not sufficient for running a systems since you have to detect discrepancies even outside of your reference settings, i.e. by observing logs, connection tests and other information sources.

The gap between both worlds is still not closed. Monitoring and management tools offer capabilities to detect changes and to do changes. Combining both means to enable the system to handle errors. Since this is done by humans, except for simple service or node restarts which can be done automatically, the increasing number of nodes and services will be an obstacle. But if the system can detect errors and react on them in a predefined manner, the system could reduce the human interactions to not yet known problems and the administrator itself has more time to solve the complex issues. For example in case of a crashed file server the clients don't need to remount their file systems and wait until the file server is ready again. If a error handling framework is flexible enough you can use this mechanism also to reduce the escalated events (in this case to single event of the crashed file server) and it is easier to find the reason very soon. Later we will describe a solution which allows fully automated error handling which offers the possibility to reduce the downtime of systems nearly to the monitoring interval of the corresponding sensors. You can also configure the system to solve only simple errors with a minimum error sensitivity and hold back other actions until an administrator is present who can react immediately on upcoming side effects.

2.5 Scheduler Awareness

The above described requirements have in common that they can be satisfied easier if the administration framework is aware of scheduling. If the monitoring is able to execute a number of tests simultaneously and the current node-to-job assignment is known, it is possible to reduce the jitter effect by synchronously executing these tests on the nodes of one job. For error handling it might also be very helpful to know the assignments to check if an error is caused by a user. Obviously it would be possible to configure the scheduler no longer to use specific named nodes for queues. Instead you might configure only the number of nodes of a certain type so that the system can force this even in case of node failures. So it seems that a connection to the scheduler would beneficial but not required.

2.6 Tool Integration

Since most providers already use some monitoring and management tools which they have adjusted to work in their specific environment, it is very hard to convince them to change to new tools. There are doubts, whether the new tools will fulfill all needs and will be able to setup the whole system at one time-effort, what prevents easy migration. Additionally, people interacting with the system prefer to work with the existing tools—this human factor should not be underestimated. It seems to be appropriate to design the new framework to allow easy integration of existing monitoring and management tools. This would have another beneficial effect—you don't have to solve their immanent problems again.

3 TIMaCS—The Project

The project TIMaCS (Tools for Intelligent System Management of Very Large Computing Systems) is initiated to solve the above mentioned issues especially for HLRS, since we will reach the point where we can't handle new resources (in quantity and quality) with the currently existing tools. TIMaCS deals with the challenges in the administrative domain upcoming due to the increasing complexity of computing systems especially of computing resources with performance of several petaflops.

The project aims at reducing the complexity of the manual administration of computing systems by realizing a framework for intelligent Management of even very large computing systems based on technologies for virtualization, knowledge-based analysis and validation of collected information, definition of metrics and policies. This framework should be able to automatically start predefined actions additionally to the notification of an administrator. Beyond that the data analysis based on previous monitoring data, regression tests and intense regular check aims at preventive actions prior to failures. The framework to be realized will include open interfaces to be easily bind to relevant existing systems like accounting or user management systems. We seek for developing a framework ready for production and their validation at the High Performance Computing Center Stuttgart (HLRS), The Center for Information Services and High Performance Computing (ZIH) and the Computing Center at the Philipps-Universität Marburg. NEC with the European High Performance Computing Technology Center and science + computing are the other partners within TIMaCS project. The project funded by the German Federal Ministry of Education and Research started in January 2009 and will end in December 2011.

3.1 Idea and Objectives

The main focus of the project is design and development of large scaling framework for monitoring and management of HPC resources. The other objectives, independent of their importance, are subordinated, since the increasing system size is even difficult to handle if you want to run the systems as usual and the next resources will be larger. The goals of the TIMaCS project are in particular:

- Concept and Implementation of a robust and highly scalable monitoring solution for very large computing systems based on existing tools and supplementary implementations ready for production.
- Design and Implementation of a system for partitioning and dynamic user assignment of very large computing systems based on concepts for virtualization. Easy setup or removal of single compute nodes out of a heterogeneous or hybrid system will be included.

- On top of that a management framework will be developed, which supports different atomization and escalation strategies based on policies, including notification of an administrator, semi-automatic to fully-automatic counteractions, prognoses, anomaly detection and their validation under production conditions.
- Development of tools for error detection and automatic error handling, as well as concepts and realization of preventive actions to check preventively the status of the infrastructure i.e. between jobs and supporting regular maintenance.
- Sustainability by defining standard conform interfaces and an integrated framework targeting at the combination of not yet synchronized developments of tools for monitoring and management, in cluster virtualization, policy based management and knowledge based data analysis.

3.2 Issues—Addressed and Not Addressed

Based on these objectives we designed a framework which allows to cover a broad variety of administrational task in providing HPC resources. The proposed framework will offer:

- Scalable monitoring and management solution for extremely large number of nodes even beyond 100.000 nodes.
- Error handling capabilities by executing predefined course of actions whenever an error is detected, either by threshold violation or by thrown error message.
- Plug in concept to integrate existing monitoring tools.
- Creation of combined sensor information like aggregated values for groups (i.e. services of a specific kind fail on one percent of all nodes) with a list of groups attached or abstract sensors like all sensors on a node related to a specific issue (i.e. network) work fine. Therefore it is easier to provide an overview over a huge system without showing to many information in detail.

We don't want to create a framework that solves all problems but we mainly look for a practicable solution ready to work in production environments. Therefore we have to narrow our framework and will not support directly:

- Topology detection since there are tools available and the gathered data only has to be adjusted. This could be probably done by simply creating some SQL statements.
- Package dependencies because operating systems bring one with them which normally can be access by scripts to read information and execute commands.
- Rule sets for error handling. Since any system has its own characteristics it would be at least very hard to set up some rules which work on every HPC resource. Therefore we will not deliver any rule set except for some testing scenarios.

3.3 Architecture

The description of the TIMaCS Architecture provided in this section is based on the description provided by the authors in an earlier paper [8]. The approach as used in TIMaCS architecture follows IBM's autonomic computing architecture, where self-managing capabilities in computer systems perform tasks that administrators chose to delegate to the technology according to predefined policies and rules. Policies are any type of formal behavioral guide prescribing the system behavior [9]. Policies can be expressed by a set of objective statements prescribing the system behavior on high level, or by a set of (event, condition, action) rules defining actions to execute in case of error-event occurrences and thus prescribing the system behavior on a low level. Policies determine the type of decisions and actions that autonomic capabilities perform. In contrast to IBM autonomic computing reference architecture, whose main purpose is to support IT professionals of server infrastructures for office environments, is the purpose of TIMaCS framework to support administrators of HPC infrastructures.

TIMaCS framework is designed as a policy based monitoring and management framework with an open architecture and hierarchical structure. The hierarchy is formed by manageability layers acting on different levels of information abstraction. This is achieved by generating state information for groups of different granularity: resources/node, node-group, cluster, organization. These granularities form abstraction layers. A possible realization of the hierarchies can be achieved in a tree-like structure, as shown in the Fig. 3.

Each manageability layer consists of dedicated nodes, with monitoring and management capabilities implemented as blocks, called MM-Node. The Monitoring block collects information from the nodes of the underlying layer, aggregates information and concludes abstraction by pre-analyzing information, creating group states of certain granularity and triggering events, indicating possible errors. In the next step, the management block (of the same MM-Node) analyzes triggered events and determines which of them needs to be investigated. Decisions are made in accordance with the predefined policies and rules, which are stored in a knowledge base filled up by system administrators when configuring framework and contains policies and rules as well as information about the infrastructure. Decisions result in actions or commands, which are submitted to and executed on managed resources (computing nodes) or other components influencing managed resources (e.g. scheduler can remove failure nodes from the batch queue). All nodes are connected by message based communication infrastructure with fault tolerance capabilities and mechanisms ensuring delivery of messages, e.g. following AMQP [10] standard.

The bottom layer, called resource/node layer, contains resources or computing nodes with integrated sensors, which provide monitoring information about resources or services running on them. Additionally, each managed resource has integrated "delegates" interfaces, which allow to execute commands on managed resources. Furthermore, there exists "delegates" which are not directly integrated in resources (e.g. Scheduler), but have indirect possibility to influence those resources (e.g. by removing error nodes from the batch queue).

The next layer, called node-group layer, collects monitoring data from several computing nodes in push or pull manner, depending on desired configuration. In order to allow integration of various existing monitoring tools (like Ganglia [11] or Nagios [5]) or other external data-sources, we use a plug-in-based concept, which allows the design of customized plug-ins, capable to collect information from any data-source. The collected data is then pre-processed, in order to generate events, which provide abstraction from the monitoring-data. The generated event is passed to the management block, where it is analyzed in order to determine whether the event requires further actions. The analysis of the event comprises furthermore operations aiming at reducing the amount of relevant events, e.g. by providing the origin reason of error event. This can be done by analyzing dependencies between events, based on predefined event relationship models or dependencies between sensors, services and hardware. After the event has been classified as error-event, it will be handed over to the policy based decision module, which makes decision on how to react on occurred error. After the decision on how to counteract an error event has been made, it is transformed into commands, which are sent to the "delegates" of managed resources (in the resource layer) or to other delegate which can influence affected resources (e.g. remove affected node from the batch queue). Each decision is reported to the upper layer, in order to provide possibility for the upper layer to intervene on taken decision, as it has also information about other parts of the cluster. The report contains description of detected error event and actions executed to handle this event. The upper layer is now able to intervene on received report by

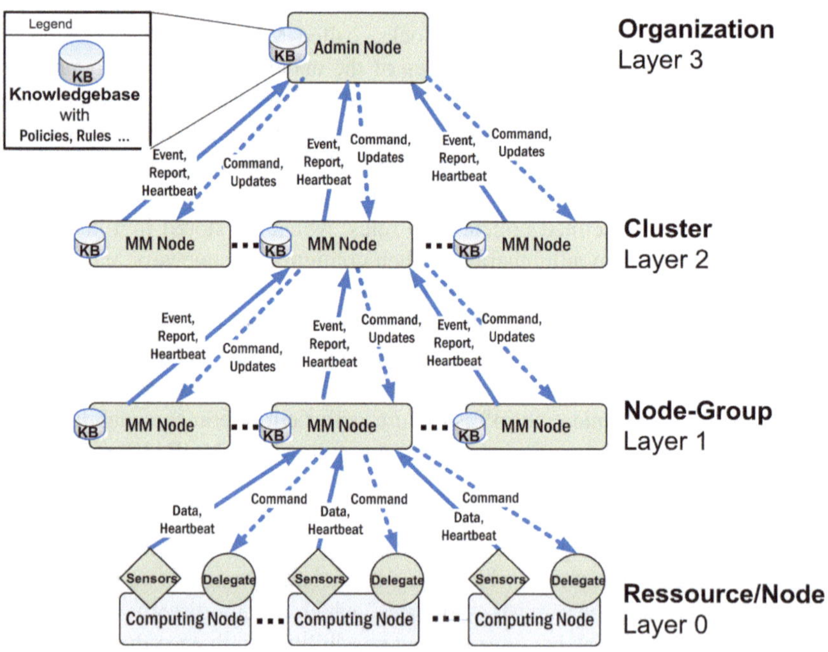

Fig. 3 TIMaCS architecture

updating the rules handling the error on the lower layer, or by executing actions affecting the managed resource. The abstraction level achieved on the group-node layer reflects the operational state of nodes or group of nodes.

The next layer, cluster layer, analyzes reports or events mediated or generated by the group-node layer and takes decision to handle escalated reports or events, e.g. dependent on severity or amount of reports received within certain time period. The analysis and evaluation of events/reports received from the group-node layer, and decision making, corresponds in principle to the description provided in node-group layer. But, in contrast to group layer, the cluster layer evaluates operational state of underlying node-groups, relating them to each other and to services and information sources needed for the operation of the whole cluster. The actions taken on this level could comprise, in addition to intervene actions, removal of affected nodes from the resource-manager, restart of jobs, rescheduling, etc.. The abstract level provided on this layer reflects the operational state of the cluster.

The top layer, organization layer, receives reports/events from several clusters of the cluster layer and takes decision on escalated events/reports. The analysis and evaluation of events/reports received, and decisions made, corresponds in principle to the description provided in cluster layer. But, in contrast to cluster-layer, the organization layer evaluates operational state of underlying clusters, relating them with other information sources, like accounting or SLA-management, and services, needed for the operation of the whole computing centre. Escalated reports/events demands that an administrator starts a deeper analysis. On top of the framework an Admin-Node is settled, which allows administrators to configure the framework, the infrastructure monitoring, to maintain the knowledge base and executing other administrative actions.

From the system monitoring point of view, the proposed framework reduces the overwhelming information flow of monitoring-data by handling and filtering it on different level of abstractions. At the same time, it increases the information value delivered to system administrator, comprising only necessary and important information. The plug-in based monitoring concept allows integration of various existing monitoring tools like Ganglia, Nagios, ZenossCore and other information sources. The collection of monitoring information can be done in push or pull manner, depending on the desired configuration of the framework. The advantage of the pull model is that collecting of monitoring information from the nodes of the underlying layer can be triggered at certain time-point explicitly. This eases control over the jitter problem by determining the time-point of information collection, e.g. between two jobs and not during the intensive activity on compute nodes.

From the system management point of view, the decision making and handling of errors is done automatically according to predefined rules and policies. This reduces system recovery time after errors and offers maximum degree of deterministic system behavior ensuring that administrators are able to retrace the system's decisions. At the same time, deterministic decision making increases administrator's confidence into the framework and helps to establish autonomic management systems in HPC. The capability of each layer to make decisions autonomously and immediately, on occurred events, increases dynamism of the system. In case of network interruptions between two layers, the lower layer is still capable to make decisions and handle occurred events independent of the upper layer. At the same time, escalated reports sent to upper layers, provide the possibility to intervene on decisions made by lower layers, without necessity to handle each report. This is achieved by smart escalation strategies.

4 Administrational Benefits

Assuming that the above described solution is ready to use, this chapter provides benefits on administration of such system and outlines influence on management of HPC resources, including simplification of system configuration.

4.1 Security Constraints

The first part handles about scenarios where security constraints are needed. Especially in case of commercial use this might be relevant to avoid interferences on access between different user groups. Normally access right on files are already in place but other users may still try to access nodes used by other users and decrease the available cpu-time for the permitted user. By preventing this i.e. on switch level, the nodes don't have to handle these attacks. Examples for security scenarios are as following:

- Access to license servers.
- Grant exclusive access to a storage system to a specific user group.
- Restrict access to node to all members of the same company.

They have in common, that the access restrictions may be enforced outside of the node hosting the service, so that the side effects can be minimized and possible security bugs in the nodes are even not reachable for attackers. Additionally these infrastructure changes can normally be changed without interrupting operation. Changes on the nodes itself often cause a service restart with a short downtime. So the critical systems' configuration has not to been changed and therefore errors in this configuration occur very seldom. The security constraints may also be enforced on the corresponding system but on infrastructure level a second barrier can be created and brute force attacks are minimized.

4.2 Mission-Critical Constraints

More important than security constraints may be mission critical issues. Especially in urgent computing scenarios it is necessary that the processes are not disturbed in any way. This can be done by:

- Encapsulating desired nodes and servers so that they appear as a separate cluster with exclusive access. This might be helpful to convince people to use even resources together with other users.
- Guarantee a certain bandwidth to an external server by restricting bandwidth for all other users on the network level.

4.3 Performance Tuning

The HPC provider itself may use the management capabilities to optimize his systems depending on currently submitted jobs, with different requirements on quality and quantity on resource usage, and users. Distributing the anticipated server workload over several servers equally, can be done by changing the server names responsible for each node. So the storage frontend which is accessed by each node, may be changed depending on the users job history or storage usage.

4.4 Dynamic Changes

Current situation in HPC provisioning is that resources are configured very static and changes need a long time to be done. With a powerful management framework this could be replaced by dynamic configurations:

- Adapt the scheduling system according to the current submission behavior. So it might be possible to increase queue size before weekends or reduce the number of resources assigned for short or interactive jobs on weekends. As a result, the number of idle nodes will decrease.
- If for performance issues some services are only available for a part of the nodes in case of permanent errors, these assignments have to be changed so that all nodes have access again. This is very important if there is a fully automated error handling in place since then you need to do changes on large systems.

4.5 Hardware Management

Even the hardware management will be easier since network hardware may only be labeled with the predefined types and the management system then knows the proper reference state which will then be enforced. So changes due to hardware failures as well as extensions to clusters may be integrated very fast.

Most of the above mentioned examples share a common idea, building virtual resources for some time which will act as they were independent.

5 Business-Benefits with Business-Policy Based Management

Policy based management approaches had been developed in order to achieve more dynamic behavior of the managed system. The hierarchical policy based management approach presented previously, allows, by definition of policies and rules on different levels of abstraction, adaptation of the managed system to different situations (i.e. errors) and dynamically changing environment—where the IT infrastructure is continuously changed and updated. In addition to different situations and dynamic IT infrastructures there are on top of the policy hierarchy different business objectives and business policies, which define the business course of the provider and influence policies on subsidiary levels. In this chapter we outline the need for business policies and show a possible approach investigating relationship between business policies and scheduling policies.

5.1 Need for Business-Policy Based Job-Scheduling in HPC

Business policies are any type of formal or informal behavioral guide prescribing behavior of/in a company that business wants to have. The behavior of high performance computing (HPC) provider can be assessed according to scheduling of jobs, with various requirements on HPC resources and Quality of Services (QoS), submitted by different users. Scheduling behavior is typically defined in a way that it implicitly adheres to business policies of HPC provider, while taking users' job requirements, available resources, existing SLAs, long term contracts and other factors into account. The range of existing schedulers used for job scheduling in HPC varies from time based scheduler like Cron [12] to advanced schedulers like Moab [13], which supports large array of scheduling policies with dynamic priorities, extensive reservations, and fair share [14]. However, schedulers have a big amount of parameters and different scheduling policies which need to be selected and adjusted in order to meet business policies in different situations.

The problem which occurs when configuring schedulers is that business policies exist in most cases implicitly. Administrators are not really aware of existing business policies and therefore configure schedulers intuitively and possibly subjectively. This makes it hard for business people to assess whether the actual scheduling behavior corresponds to current business policies, as it requires understanding of scheduling configuration parameters.

Additionally, there might be a fast switch required between different business policies, dependent on the occurred situation. For instance, in profit oriented organizations, managers try to achieve maximum return on investment which often means that they only deliver various qualities of services to various users and groups [14] to increase system utilization. In contrast, nonprofit organizations, like national computing centres, have their focus on delivering various qualities of services to various (or certain) users and group even if this will cause a decreasing utilization. Some of the national computing centres have joint collaboration with scientific and industrial partners through common joint cooperation company. That means scheduling behavior in clusters of such computing centres needs to be flexible enough to be adapted to various business needs even at the same time.

5.2 Approach

As outlined, there is a need for rapid changes of the scheduling behavior in order to adapt it to changing business requirements (from "high system utilization" to "customer satisfaction") and new situations. An approach to handle such kinds of problems, induced by changing business objectives and altering situations, might follow IBM's autonomic computing reference architecture [9]. Autonomic computing is thereby defined "as a computing environment with the ability to manage itself and dynamically adapt to change in accordance with business policies and objectives" [9]. Following this approach, there must be business policies defined, capable to express business requirements influencing scheduling behavior on high level. In contrast to IBM autonomic computing reference architecture is the purpose of the proposed approach to support HPC providers on business level in job-scheduling.

The investigation on the relationship between business policies and scheduling policies can be achieved by developing a model, capable to map HPC business policies together with SLAs, long term contracts, available resources and other elements influencing scheduling behavior, to selection and adjustment of scheduling policies and configuration parameters, used to configure scheduler and thus to provide the scheduling behavior in accordance with the business policy. Thereby an intermediate step is necessary, which requires development of HPC business policy specification, capable to capture and describe behavior of HPC centre on high level.

The approach follows a bottom-up process: The first step thereby is to analyze existing scheduling policies, in order to identify most common key factors (with their interrelation and possible hierarchy), like priority of users/jobs, fairness, response-time, utilization-strategy, available resources, SLAs/contracts, and other factors influencing scheduling behavior. The next step consists in the analysis of existing business policies and their relationship to identified key factors in different situations. The outcome of the second step will be a model, which explains relationships between current business policies, the key factors and scheduling policies. The third step consists of the identification of HPC business policy schema, derived from the model in second step, capable to express HPC business policies. Finally, in order to evaluate results achieved in previously steps, the last step consists of reference implementation, enabling mapping of reference business policies together with other key factors to scheduling policy configuration for advanced schedulers like Moab [13] or Maui [14].

6 Conclusion and Future Outlook

We have shown that for future HPC resources it is necessary to develop a new generation of monitoring and management framework, since existing tools are not capable to cover all HPC relevant aspects, because of scalability problem and other HPC related issues. Especially the difficulty in avoiding additional jitter effects is increasing with the size of resources. The proposed framework within the TIMaCS project is capable of fulfilling most of the requirements we have identified for productional use.

Assuming the framework is developed as described, we have foreseen how this can change HPC administration from the more technical view, to help administrators to manage larger and more complex systems. From the business point of view, ability to express business policies on high level allows flexibly and fast adaptation of the system to different situation, where new type of services can be offered to customers. Beyond that, the presented work offers more possibilities in area which are only on the fringe today. If it is possible to monitor nearly everything within a resource and react on any deviations, this system could even be used for user initiated management. So they can monitor performance within their jobs and optimize the code or allocate resource accordingly, to increase i.e. calculated details. Or they can catch errors within the jobs and create error handling strategies reducing i.e. jitter problem.

We think that the management of very large resources will be changed dramatically in the next years, since current procedures are to static to satisfy future requirements. This might even result in a paradigm change for resource management.

References

1. TIMaCS project web-site http://www.timacs.de
2. Top 500 supercomputing Sites http://www.top500.org
3. Rami Matarneh (2009). Multi Microkernel Operating Systems for Multi-Core Processors, *Journal of Computing Science* 5 (7) (pp. 493–500). ISSN 1549-3936
4. Linux Magazine, Technical Review, Monitoring (2007)
5. Nagios project web-site http://www.nagios.org. Cited 28 May 2010
6. Big Brother product web-site http://www.bb4.com
7. Zenoss project web-site http://www.zenoss.com
8. Buchholz, J., Volk, E.: The Need for New Monitoring and Management Technologies in Large Scale Computing Systems. In: Proceedings of eChallenges 2010, to appear
9. IBM: An architectural blueprint for autonomic computing http://www-01.ibm.com/software/tivoli/autonomic/pdfs/AC_Blueprint_White_Paper_4th.pdf, IBM Whitepaper, June 2006. Cited 28 May 2010

10. AMQP web-site http://www.amqp.org. Cited 28 May 2010
11. Ganglia sourceforge web-site http://ganglia.sourceforge.net. Cited 28 May 2010
12. Wikipedia: Cron description http://en.wikipedia.org/wiki/Cron. Cited 28 May 2010
13. Clusterresources: Moab Workload Manager user-guide http://www.clusterresources.com/products/mwm/docs/moabusers.shtml. Cited 28 May 2010
14. Clusterresources: Maui Scheduler Administrator's Guide, version 3.2 http://www.clusterresources.com/products/maui/docs/. Cited 28 May 2010

Empirical Optimization of Collective Communications with ADCL

Katharina Benkert, Edgar Gabriel

Abstract The Abstract Data and Communication Library (ADCL) allows for auto-tuning of communication operations for parallel applications. This paper presents a new set of interfaces introduced in ADCL in order to support most MPI collective communication operations, and thus enable the optimization of one of the most widely used features of the MPI specification. The paper discusses semantic as well as implementation aspects, and evaluates the new interfaces using the NPB FT benchmark on a large selection of platforms and MPI libraries.

1 Introduction and Motivation

Automatic performance tuning is an area of research defined by one of the fundamental questions in computing: how to provide with reasonable effort efficient code on a wide variety of computer architectures steadily increasing in complexity and diversity. Or, in other words, how to obtain for a kernel (computational kernel, communication pattern) platform-independently an equal or superior performance compared to hand-tuned code.

Collective operations are frequently used. Rabenseifner showed in a long-term study in [9] that apart from point-to-point communications a significant amount of communication time is spent in the collectives `MPI_Alltoall` and `MPI_Allreduce`. Terry Jones presented in a more recent study [8] that the time

Katharina Benkert
High Performance Computing Center Stuttgart (HLRS), University of Stuttgart, 70550 Stuttgart, Germany
e-mail: benkert@hlrs.de

Edgar Gabriel
Parallel Software Technologies Laboratory, Department of Computer Science, University of Houston, Houston, TX, USA
e-mail: gabriel@cs.uh.edu

M. Resch et al. (eds.), *High Performance Computing on Vector Systems 2010*,
DOI 10.1007/978-3-642-11851-7_3,

spent in collectives is predominantly divided among barrier, allreduce, broadcast, gather and alltoall and that the alltoall performance is vital to some codes.

Since the performance of collective operations has a great influence on the scalability, many researchers tried to improve the efficiency of collectives under certain assumptions on the type of network and the message length. However, no single algorithm can lead to optimal performance in all possible scenarios. Even if the MPI implementation combines several algorithms using a heuristic switching technique depending on the message length, the best time to solution is not guaranteed. Letting a common HPC user try to figure out optimization possibilities and system sensitivity, by having him implement several communication possibilities and run extensive benchmarks with different problem sizes, number of processes and MPI libraries on different machines, is no solution either.

An answer to this dilemma are empirical optimization libraries. They possess in general three distinct components: a large number of different code-lets for the kernel that is being optimized, timing routines to measure the execution time of the code-lets empirically, i.e. by actually running the code-lets, and selection algorithms to determine the fastest code-let based on the obtained empirical data.

The Abstract Data and Communication Library (ADCL) is an auto-tuning library which acts as an add-on to the MPI implementation. It intends to take away complexity from the user and at the same time exploits optimization possibilities. In [6] the benefits of using ADCL for a specific communication pattern—the n-dimensional Cartesian neighborhood communication which is commonly used in applications with stencil computations—were demonstrated.

In this paper we describe enhancements to the ADCL API in order to extend ADCL's functionality beyond Cartesian neighborhood communication, most notably to support MPI collective operations. The main challenge lies in the fact that up to now data description and information about the communication operation are intertwined. The main contribution of this paper therefore is to develop and detail a new set of interfaces supporting more generic communication operations. We implement an ADCL version of the NAS FT benchmark using the new interfaces for the all-to-all communication pattern and present the results for various problem sizes, number of processors and MPI libraries on several HPC systems.

2 The Abstract Data and Communication Library (ADCL)

The Abstract Data and Communication Library (ADCL) [5] is an empirical auto-tuning library targeting mostly MPI communications, although its functionality could be used to optimize any user-provided set of code-lets. It aims to provide the highest possible performance for application level communication patterns within a given execution environment. At present, the library has a predefined set of 20 code-lets for the n-dimensional Cartesian neighborhood communication and incorporates a run-time selection and decision logic in order to choose the code-let leading to the highest performance.

ADCL has its own API since its goal is to tune any code fragment. In case of complicated communication patterns with more than one MPI communication, it becomes difficult to impossible to use the performance interface of MPI or to implement a compiler-based automatic code substitution. After an MPI program is started, the performance of a MPI data exchange is regarded a function of (the size of) the data, the communication pattern and the process topology. For each of this parameters exists a corresponding ADCL object:

- `ADCL_Vector` describes the data to be communicated. The object contains the dimensions and extents of the data array, its basic data type, the number of components per grid point *nc*, the number of layers of halo cells *hwidth* and a pointer to the data array.
- `ADCL_Function` is a code-let that implements a certain numerical kernel or communication pattern.
- `ADCL_Fnctset` is a set of ADCL functions providing the same functionality. ADCL includes a pre-defined function set `ADCL_FNCTSET_NEIGHBORHOOD` for the *n*-dimensional Cartesian neighborhood communication.
- `ADCL_Topology` provides a description of the process topology and neighborhood relations within the application.
- `ADCL_Request` combines a vector object, a function set and a topology object. An `ADCL_Request` represents a persistent communication object, which similarly to its MPI counterpart for sequential persistent requests can be 'started', in this case using `ADCL_Request_start`.

The following code sample gives a simple example for an ADCL code, using a 2D neighborhood communication on a 2D process topology.

```
int ndim = 2;             /* number of dimensions */
double **vector;          /* data array with halos to be communicated */
int vec_dims = [7,5];     /* extents of the data array */
int nc       = 1;         /* entries per grid point */
int hwidth   = 1;         /* number of layers of halo cells */

ADCL_Vector vec;
ADCL_Topology topo;
ADCL_Request request;

/* Allocate a 2D vector with ADCL */
ADCL_Vector_allocate (ndims, vec_dims, nc, ADCL_VECTOR_HALO, hwidth,
     MPI_DOUBLE, vector, &vec);

/* Generate a 2-D process topology */
MPI_Cart_create (MPI_COMM_WORLD, 2, cart_dims, periods, 0, &cart_comm);
ADCL_Topology_create (cart_comm, &topo );

/* Combine description of data structure, predefined function set and
   process topology */
ADCL_Request_create (vec, topo, ADCL_FNCTSET_NEIGHBORHOOD, &request );
```

```
/* Main application loop */
for (i=0; i<NIT; i++ ) {
    ...
    /* Initiate neighborhood communication */
    ADCL_Request_start (request );
    ...
}

ADCL_Request_free   ( &request );
ADCL_Topology_free  ( &topo );
ADCL_Vector_free    ( &vec );
```

`ADCL_Request_start` replaces the calls to MPI. They execute the communication and control the empirical optimization. During the first iterations of the application, the fastest code-let from the neighborhood function set is determined: execution times of alternative code-lets are measured multiple times one after another. After all code-lets have been tested the required number of times, the measurements are analyzed mainly locally with a statistical method as explained in [2]. Then, depending on the evaluation method used, the method judged fastest is chosen and used during the remainder of the simulation.

3 Semantics of the New ADCL Interfaces

The ADCL vector object as presented in Sect. 2 serves two purposes: first, it allows to identify the buffer associated with a communication operation; second, it allows to perform an automatic data mapping of which portion of the data array is supposed to be transferred to which process. As an example for the latter the interface to allocate an ADCL vector,

```
int ADCL_Vector_allocate ( int ndims, int *dims, int nc, int comtype,
    int hwidth, MPI_Datatype dat, void *data, ADCL_Vector *vec )
```

contains the parameter `hwidth` which specifies the number of layers of halo cells that have to be transferred to the neighboring processes. Combined with the topology object described previously, the library automatically determines which elements have to be transferred to which process.

3.1 The Vector-Map Object

Although this functionality is highly convenient, the main restriction of this API was its limiting the functionality to Cartesian neighborhood communication, the original driving force of the library. To support further communication patterns, we developed therefore a set of new API interfaces, which allow to separate the description of the communication buffer and the mapping of which elements of the buffer have to be transferred to which process.

The new set of interfaces developed within this project tries to accommodate multiple goals:

1. allow for the definition of user defined functions as well as predefined operations
2. separate data management from the actual communication operations
3. allow for a light-weight description of data mappings and the automatic association with remote processes.

Collective	Data information	Communication information	Topology information	vmap type for svec (& rvec)
MPI_Bcast	buffer, datatype	count	root, comm	all
MPI_Gather	sbuf, stype, rbuf, rtype	scount, rcount	root, comm	all & all or inplace & all
MPI_Gatherv	sbuf, stype, rbuf, rtype	scount, rcounts, displs	root,comm	all & list or inplace & list
MPI_Scatter	sbuf, stype, rbuf, rtype	scount, rcount	root, comm	all & all or inplace & all
MPI_Scatterv	sbuf, stype, rbuf, rtype	scounts, displs, rcount	root, comm	list & all or inplace & list
MPI_Allgather	sbuf, stype, rbuf, rtype	scount, rcount,	comm	all & all or inplace & all
MPI_Allgatherv	sbuf, stype, rbuf, rtype	scount, rcounts, displs	comm	all & list or inplace & list
MPI_Alltoall	sbuf, stype, rbuf, rtype	scount, rcount	comm	all & all or inplace & all
MPI_Alltoallv	sbuf, stype, rbuf, rtype	scounts, sdispls, rcounts, rdispls	comm	list & list or inplace & list
MPI_Alltoallw	sbuf, stypes, rbuf, rtypes	scounts, sdispls, rcounts, rdispls	comm	list & list or inplace & list
MPI_Reduce	sbuf, rbuf, datatype	count, op	root, comm	reduce (2x) or inplace & reduce
MPI_Allreduce	sbuf, rbuf, datatype	count, op	comm	reduce (2x) or inplace & reduce
MPI_Reduce_Scatter_block	sbuf, rbuf, datatype	rcount, op	comm	reduce (2x) or inplace & reduce
MPI_Reduce_scatter	sbuf, rbuf, datatype	rcounts, op	comm	redscatter (2x) or inplace & redscatter
MPI_Scan	sbuf, rbuf, datatype	count, op	comm	reduce (2x) or inplace & reduce
MPI_Exscan	sbuf, rbuf, datatype	count, op	comm	reduce (2x) or inplace & reduce

Table 1 Overview of collectives. Interface parameters are divided into data, communication and topology-related information (s—send, r—recv)

The new interface therefore distinguishes between five different objects: the vector object, the vector-map object, the topology object, the function-set and the request. In the following, we would like to focus our attention to the vector, vector-map and the request object.

The main purpose of the 'new' vector object is to define a data array that will be used later for communication. The interface allows to *register* or *allocate* a multi-dimensional memory region, using the number of dimensions of the data array, extent of each array, number of elements of each data point in the array, and the basic MPI datatype. The vector does **not** specify which elements of the data array will be used in communication operations.

The vector-map object (or short 'vmap') allows to define which elements of the vector object have to be transferred to which process. This functionality does not have a counterpart in MPI, since it combines functionality often provided by the vector versions of the MPI collective operations such as `MPI_Gatherv`, `MPI_Scatterv` and derived MPI data types. Although the vmap object does not have the flexibility of the most generic MPI derived data type constructors such as `MPI_Type_create_struct`, it provides a much simpler and user-friendlier interface compared to the latter one, and covers many if not most common situations. Specifically, based on the vector and the vmap object, the ADCL library is able to construct the required derived MPI data types automatically for the end-user.

3.2 Extension of the ADCL Interfaces

In order to support a large variety of communication patterns, we analyze in the following the information required for the MPI collective operations, the Cartesian neighborhood communication and user defined function sets.

For the collectives defined in the MPI standard and shown in Table 1 we notice that the parameters can be separated into three groups: information concerning the data (buffer, data type) which is stored in the ADCL vector object, information concerning the process topology (communicator, root) stored in the ADCL topology object and finally information related to the communication pattern (element counts, reduction operation, array of element counts or displacements) which becomes part of the new vmap object. If one treats send and receive information separately, one obtains four different types of vmaps for the collectives. Together with an `inplace` type and the `halo` type for Cartesian neighborhood communication, this results in six different types of vmap objects. The constant `ADCL_VMAP_NULL` can be used for user-defined function sets that do not necessarily need a vmap object. Table 2 summarizes the different vector-map object types and the required information for each of them.

Due to the broad range of parameters required for various operations, different interfaces for different operations have been defined. The parameter `comtype` was originally used in the interfaces to allocate or register a vector and describes the type of vmap. It now becomes part of the interface to allocate the vmap object as `vmap_`*comtype*`_allocate`. The new interfaces for vmap, vector and request creation for the example code from Sect. 2 are now

Vmap comptype	Parameters
inplace	–
halo	hwidth
all	count
reduce	count, op
list	counts, displs
redscatter*	counts, op

Table 2 Types of ADCL vmap objects (*—not implemented as only needed for MPI_Reduce_scatter)

```
ADCL_Vmap_halo_allocate ( int hwidth, ADCL_Vmap *vmap );
ADCL_Vector_allocate ( int ndims, int *dims, int nc, MPI_Datatype dat,
                       void *data, ADCL_Vector *vec )
...
ADCL_Request_create ( ADCL_Vector vec, ADCL_Vmap vmap,
                      ADCL_Topology topo, ADCL_FNCTSET_NEIGHBORHOOD,
                      ADCL_Request *request);
```

Special attention has to be attributed to the data types which naturally belong to the vector object. However, also the request object has data types since in case of the neighborhood communication, the vector object contains the basic data type whereas for the request derived types are constructed which depend on the topology information (size), data information (dimensions and extent of the data array, *nc*) and communication information (hwidth). This means that for collectives the data types from the vector object have to be copied to the request object. New variables are introduced in the request object which specify the number of MPI data types to be sent or received. For the copy operations of the data types and the initialization of the new variables, the functions ADCL_basic_init and ADCL_basic_free are implemented.

3.3 The New Function Sets

For supporting MPI collective operations within ADCL, we encapsulate the native MPI collective provided by the MPI library as one code-let. Additionally, we implement a variety of algorithms which perform the collective communication based on pairwise communications. Inside the MPI library, the collective function is likewise performed by one algorithm as a sequence of pairwise communications, but it is up to the MPI library which algorithm it uses and in case of vendor MPI libraries not known to the user.

New predefined function sets for allreduce, allgatherv and all-to-all have been added. ADCL provides 5 code-lets for the allreduce operation. One is the encapsulated native `MPI_Allreduce`, the 4 others are based on send-receive operations: linear (reduce to root and broadcast with own implementations), non-overlapping (reduce and broadcast with MPI implementation), recursive doubling algorithm as used in MPICH2 [7] for small and intermediate size messages. and ring.

For allgatherv, there is the native MPI implementation, linear (gatherv to root and broadcast), recursive-doubling as used in MPICH2 [7], Bruck (a variation of the All-to-all algorithm described in [3]), neighbor exchange (adapted from allgather algorithm described by Chen et.al. in [4]) and ring.

The all-to-all function set consists of eight code-lets, here numbered for reference in Sect. 4: the native `MPI_all-to-all` (C0), linear_sync (C1), pairwise (C2), pairwise_excl (C3), linear (C4) and Bruck's Algorithm [3] with a minor modification as used in MPICH2 [7], which restricts the number of messages. In our case we use block sizes of 2, 4 and 8 (C4–C7).

No special effort has been invested to tune the different code-lets as of today. In a long-term, we plan to add the flexibility for supporting various data transfer primitives (blocking, non-blocking, one-sided) and various methods to handle non-contiguous data, similarly to the Cartesian neighborhood communication.

4 Performance Evaluation

To compare the code-lets of ADCL with the native MPI_Alltoall, we use the MPI FFT Benchmark of the NAS Parallel Benchmarks 3.0 [1] from NASA Ames Research Center. The CFD related "paper and pencil" benchmarks consist of a set of 5 computational kernels (EP, MG, CG, FT, and IS) and 3 pseudo applications (LU, SP and BT) and serve to evaluate supercomputers. Each of the kernels addresses a different type of numerical computation. The FT benchmark computes the solution of a 3-dimensional partial differential equation with Fast Fourier Transforms (FFTs). FFTs are often used in spectral methods and for large Eddy turbulence simulations and require all-to-all communications for matrix transpose operations. Consequently, the FT benchmark creates substantial communication and evaluates network performance.

4.1 Integration of ADCL

The ADCL implementation is based on a 1-dimensional data distribution. In the main program, after the call to the benchmark's setup() routine, ADCL_Init is called and the ADCL data structures are build: a vmap object for the all-to-all communication, vector objects for u_1 and u_2 are registered and a request object is allocated. A switch is set to false to avoid counting the first call to fft(). It is set to true right before the main loop of the FT benchmark which consists of an evolution step and the computation of the FFT. During its execution in the subroutine transpose2_global the call to MPI_Alltoall is replaced by an ADCL_Request_start. After the call to print_timers() the ADCL objects are deregistered or deallocated and ADCL_Finalize is called. The header file ADCL.inc is included in the main program as well as in the subroutine.

```
 program ft
 include 'ADCL.inc'
c  FT: further includes and declarations
  call MPI_Init(ierr)
  call ADCL_Init(ierr)
c FT: timing and setup

c set up ADCL data structures
  call adcl_topology_create ( MPI_COMM_WORLD, adcl_topo, ierr )
  call adcl_vmap_alltoall_allocate( ntdivnp/np, ntdivnp/np,
 >     adcl_vmap, ierr )
  call adcl_vector_register ( 1, ntdivnp, 0, dc_type, u2,
 >     adcl_svec, ierr )  ! send vector
  call adcl_vector_register ( 1, ntdivnp, 0, dc_type, u1,
 >     adcl_rvec, ierr )  ! receive vector
  call adcl_request_create_generic ( adcl_svec, adcl_vmap,
 >     adcl_rvec, adcl_vmap, adcl_topo, ADCL_FNCTSET_ALLTOALL,
 >     adcl_request, ierr )

c disable adcl while problem is ran once for benchmarking reasons
  use_adcl = .false.
c run problem
  use_adcl = .true.

c main loop
  do iter = 1, niter
     call evolve(...)
     call fft(...) ! calls transpose_xy_z which calls transpose2_global
  end do

c FT: verification and output

c free ADCL objects
  call adcl_request_free ( adcl_request, ierr )
  call adcl_topology_free ( adcl_topo, ierr )
  call adcl_vector_deregister( adcl_svec, ierr )
  call adcl_vector_deregister( adcl_rvec, ierr )
  call adcl_vmap_free (adcl_vmap, ierr )

  call ADCL_Finalize(ierr)
  call MPI_Finalize(ierr)
  end program FT

  subroutine transpose2_global(xin, xout)
  include 'ADCL.inc'
c FT: further includes, declarations and timing
  call adcl_request_start ( adcl_request, ierr ) ! replaces
c FT:  call mpi_alltoall(xin, ntdivnp/np, dc_type,
c FT: >                  xout, ntdivnp/np, dc_type,
c FT: >                  commslice1, ierr)
c FT: timing
  end
```

4.2 Setup

The test systems used are a Nehalem cluster with InfiniBand interconnect, a Cray XT5m and a NEC-SX8 installation at HLRS, the SGI Altix at LRZ Munich and the BlueGene/P system at the Supercomputing System Jülich. For each system, we executed the FFT benchmark for various classes K with different numbers of processes n, abbreviated as nK in the figures below, and eventually with multiple MPI implementations. The dimensions of the domain as well as the message sizes for the different test cases are shown in Table 3. Runs were executed in the virtual node mode, i.e. every core ran an MPI process.

Class	nx	ny	nz	Total message size[GB]	#procs	Message size per process [KB]
A	256	256	128	0.1	8	2097
B	512	256	256	0.5	8	8388
					32	524
C	512	512	512	2.1	32	2097
					128	131
					256	32
					512	8

Table 3 Message sizes for the FFT benchmark

Within a single batch job, we execute three set of runs. Each set consists of 9 runs, one for each of the 8 code-lets presented in Sect. 3.3 for the all-to-all operation and one without ADCL, with 200 FFT iterations.

4.3 Results

We follow a two-step procedure to analyze the results. At first, we compare the execution time of the winner code-let to that of the encapsulated native MPI_Alltoall. Secondly, we take into account the overhead caused by ADCL and compare the execution time of the ADCL winner code-let to the benchmark results without ADCL.

In the following, the subscript w refers to the winner code-let, n to the encapsulated native MPI_Alltoall and o to the original implementation without ADCL. We compute the mean $\bar{t}_k = \frac{1}{N}\sum_{i=1}^{N=3} t_{k,i},\ k \in \{w,n,o\}$ from the execution times $t_{k,i}$ of the three runs. The degree of dispersion about the mean is expressed by the uncertainty $u_k = \frac{s_k}{\sqrt{N}}$ where

$$s_k = \sqrt{\frac{1}{N-1}\sum_{i=1}^{N}(t_{k,i} - \bar{t}_k)^2}$$

is the sample standard deviation.

For $k \in \{n, o\}$, the possible gains in percent are given by

$$g_k(\bar{t}_k, \bar{t}_w) = \frac{\bar{t}_k - \bar{t}_w}{\bar{t}_k} \cdot 100$$

with uncertainty

$$utot_k = \sqrt{\left(\frac{\partial g_k}{\partial t_k} \cdot u_k\right)^2 + \left(\frac{\partial g_k}{\partial t_w} \cdot u_w\right)^2}$$
$$= \sqrt{\left(\frac{t_w}{t_k^2} \cdot u_k\right)^2 + \left(\frac{u_w}{t_x}\right)^2} \cdot 100\,.$$

When comparing the winner code-let to the encapsulated native MPI_Alltoall, in 22 out of 42 test cases, i.e. more than 50%, the encapsulated all-to-all was not the best implementation as shown in Table 4. Among them were 6 cases in which ADCL performed a lot better. It is important to note that 7 out of 8 code-lets provided by ADCL performed best in at least one test case.

Test case	Winner code-let	Gain[%]	$utot_n$
nehalem_impi_8A	C7	0.67	0.19
nehalem_impi_8B	C7	2.03	0.51
nehalem_impi_128C	C2	0.05	1.07
nehalem_impi_512C	C3	**35.85**	0.87
nehalem_ompi_8A	C4	1.83	1.23
nehalem_ompi_8B	C1	0.34	0.50
nehalem_ompi_32B	C3	1.74	0.96
nehalem_ompi_32C	C3	0.46	1.88
nehalem_ompi_128C	C2	1.52	0.89
nehalem_ompi_512C	C3	**34.31**	2.21
sgi_altixmpi_256C	C2	**20.27**	2.76
sgi_impi_8A	C7	3.56	0.25
sgi_impi_8B	C4	**22.44**	8.54
sgi_impi_32B	C4	4.92	0.29
sgi_impi_128C	C4	**11.32**	1.47
sgi_impi_256C	C1	4.16	5.13
sgi_ompi_8B	C4	2.16	0.86
sgi_ompi_32B	C3	**56.23**	23.91
cray_8A	C7	0.70	0.15
cray_8B	C7	1.14	0.34
cray_128C	C3	0.38	0.71
cray_256C	C3	6.82	6.62
sx8_8A	C7	0.74	0.07
sx8_8B	C4	0.32	0.07
sx8_32B	C4	1.35	0.50
jugene_vn_8A	C5	1.84	0.01
jugene_vn_32B	C1	2.14	0.00

Table 4 Overview of testcases with other winning code-lets than the encapsulated native MPI implementation C0

Secondly, we take into account the overhead caused by ADCL and compare the execution time of the ADCL winner code-let to the results of the original benchmark without ADCL. The possible gains and losses using ADCL are depicted in Fig. 1. Except for the test cases sgi_altixmpi_8B, sgi_altixmpi_32B and jugene_vn_512C, ADCL performs as good or better than the original version without ADCL.

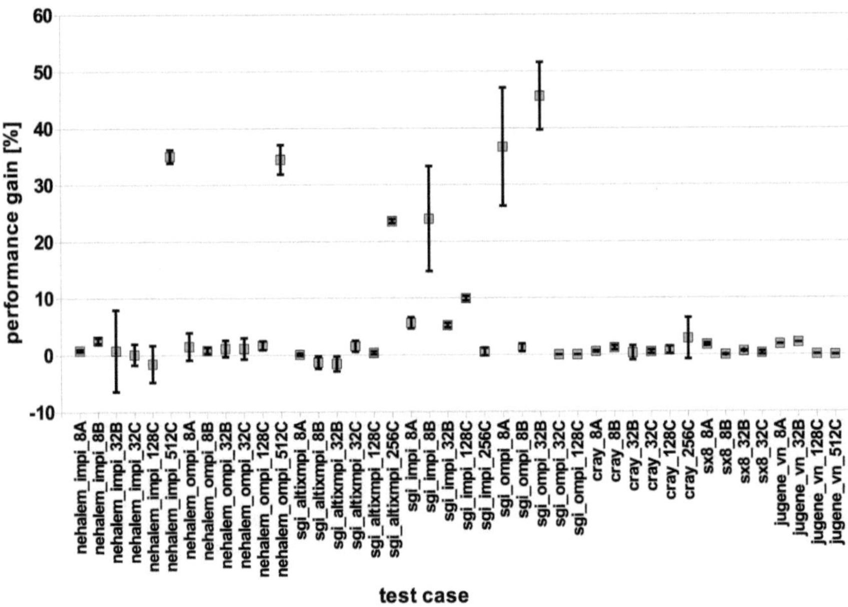

Fig. 1 Performance gains and losses in percent when comparing the winner code-let of ADCL to the original benchmark version without ADCL. The error bars show the uncertainty $utot_o$

5 Summary and Outlook

In this paper we described the goals, semantics and realization of a set of new interfaces to extend the auto-tuning library ADCL. With the recently introduced vector-map object, data management and communication information are separated to allow for new communication patterns or user-defined functions. The approach has been validated by analyzing the requirements of collective operations and user-defined functions and implementing predefined function sets for allreduce, allgatherv and alltoall.

Although the initial implementation for the predefined function sets does not support various data transfer primitives or methods to handle non-contiguous data, the results obtained for the NAS FT benchmark with its all-to-all communication pattern showed that in all but 3 cases an equal or superior performance could be achieved when using ADCL.

Acknowledgements This work was funded by the project STEDG within the BMBF Software Initiative for High Performance Computing. Partial support for this work was provided by the National Science Foundation under award no. CNS-0846002. Any opinions, findings, and conclusions or recommendations expressed in this material are those of the authors and do not necessarily reflect the views of the National Science Foundation.

References

1. Bailey, D., Barszcz, E., Barton, J., Browning, D., Carter, R., Dagum, L., Fatoohi, R., Fineberg, S., Frederickson, P., Lasinski, T., Schreiber, R., Simon, H., Venkatakrishnan, V., Weeratunga, S.: The NAS Parallel Benchmarks (1994)
2. Benkert, K., Gabriel, E., Resch, M.M.: Outlier Detection in Performance Data of Parallel Applications. In: 9th IEEE International Workshop on Parallel and Distributed Scientific and Engineering Computing (2008)
3. Bruck, J., Ho, C.T., Kipnis, S., Weathersby, D.: Efficient algorithms for all-to-all communications in multi-port message-passing systems. In: SPAA '94: Proceedings of the sixth annual ACM symposium on Parallel algorithms and architectures, pp. 298–309. ACM, New York, NY, USA (1994). DOI http://doi.acm.org/10.1145/181014.181756
4. Chen, J., Zhang, Y., Zhang, L., Yuan, W.: Performance evaluation of allgather algorithms on terascale linux cluster with fast ethernet. International Conference on High Performance Computing and Grid in Asia Pacific Region, 437–442 (2005). DOI http://doi.ieeecomputersociety.org/10.1109/HPCASIA.2005.75
5. Gabriel, E., Feki, S., Benkert, K., Resch, M.M.: Towards Performance Portability through Runtime Adaption for High Performance Computing Applications. Concurrency and Computation—Practice and Experience, accepted for publication (2010)
6. Gabriel, E., Huang, S.: Runtime optimization of application level communication patterns. In: Proceedings of the 2007 International Parallel and Distributed Processing Symposium, 12th International Workshop on High-Level Parallel Programming Models and Supportive Environments, p. 185 (2007)
7. Gropp, W., Lusk, E., Doss, N., Skjellum, A.: A high-performance, portable implementation of the MPI message passing interface standard. Parallel Computing **22**(6), 789–828 (1996)
8. Jones, T.: Survey of MPI Call Usage. In: IBM System Scientific User Group (ScicomP) 10 (2004)
9. Rabenseifner, R.: Automatic MPI Counter Profiling. In: 42nd CUG Conference. Noorwijk, The Netherlands. (2000). URL http://scholar.google.com/scholar?hl=en&btnG=Search &q=intitle:Automatic+MPI+Counter+Profiling#0

Part II
I/O Strategies

I/O Forwarding on NEC SX-9

Erich Focht, Thomas Großmann, Danny Sternkopf

Abstract I/O Forwarding can give the NEC SX-9 access to non-natively supported filesystems like Lustre. The implementation of the first IOFWD release has been finished. The IOFWD framework is running on the SX-9 as a client and on x86-64 based machines as a server. Users can benefit in several ways like running coupled applications spanning over scalar and vector machines and simplify their workflows by avoiding the hassle of copying data between various filesystems. The development was done as part of the SX-Linux project in collaboration of HLRS and NEC EHPCTC Stuttgart. This paper reports the implemented IOFWD design and the usage of IOFWD on the SX-9.

1 IOFWD Implementation

The core idea behind I/O forwarding is simple and natural: a highly sophisticated computer specialized to do floating point operations at highest possible rate, like the SX-9 vector supercomputer, should spend every cycle for what it does best: computation. It should not need to care about files, I/O, filesystems, block devices. A general purpose scalar CPU like an x86_64 running the Linux operating system is much better suited to do I/O on behalf of a SX-9 user program. Therefore I/O calls issued by a program running on the SX-9 are intercepted, transformed into RPCs that are sent to a designated I/O server. On the server side the RPCs run local I/O operations, gather the results and send them back to the SX-9. This is accomplished by the IOFWD infrastructure.

Erich Focht, Danny Sternkopf
NEC Deutschland GmbH, Hessbrühlstr. 21b, 70565 Stuttgart, Germany
e-mail: efocht@hpce.nec.com, dsternkopf@hpce.nec.com

Thomas Großmann
High Performance Computing Center Stuttgart (HLRS), Nobelstr. 19, 70569 Stuttgart, Germany
e-mail: grossmann@hlrs.de

M. Resch et al. (eds.), *High Performance Computing on Vector Systems 2010*,
DOI 10.1007/978-3-642-11851-7_4,

Physically IOFWD uses a 10 Gigabit Ethernet network to invoke the I/O operations and transfer arguments and data buffers from a SX-9 node to an I/O forwarding server running Linux on x86_64 architecture. We have implemented an user-space approach half-integrated into a C library and additionally libraries that redirect system calls to a daemon running on I/O forwarding servers.

1.1 Design of IOFWD

The design of IOFWD is based on a client-server architecture. The compute processes are running on the SX-9 and the IOFWD server daemon, also called **iofwd-srv**, is running on the I/O forwarding servers. I/O operations from the SX compute process will be forwarded to the iofwdsrv which does the actual I/O on a local or parallel filesystem.

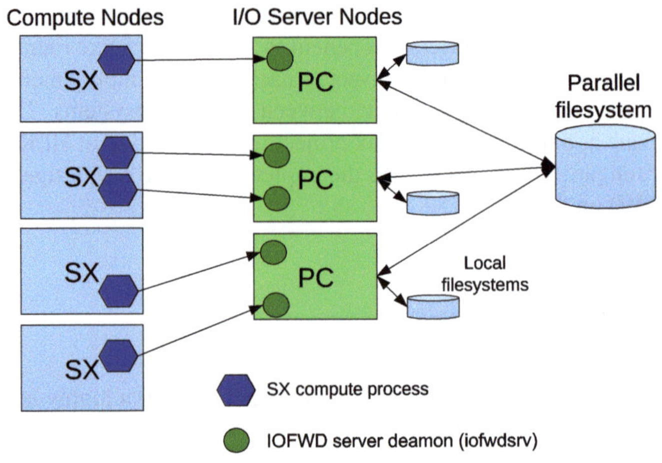

Fig. 1 IOFWD design: SX is a client, PC is a server

The iofwdsrv daemon runs on a Linux host and can access cluster/parallel filesystems through standard Linux filesystem client software. The I/O forwarding server acts as a gateway for I/O between networks. It uses 10 GE to communicate with the SX and it uses Infiniband to access the Lustre filesystem.

Each compute process on the SX spawns one **iofwdsrv** on the I/O forwarding server to establish a dedicated connection. This is sketched in Fig. 1. This approach is very flexible and allows using the same server for I/O of a parallel job e.g. when accessing a local filesystem on the server, but also allows distributing the I/O load easily to multiple servers when using a parallel filesystem like Lustre.

1.2 IOFWD Components

IOFWD is layered into several components. On the client side these are:

- A system call interception layer that replaces the I/O related calls with RPCs, serializes the call arguments and deserializes the call results.
- Libiofwd, a simple RPC library used for transferring commands, arguments and results.
- A network abstraction used by Libiofwd for transferring RPCs "on the wire".

The stack is slightly different on the server side where the layer corresponding to the syscall interceptor consists of RPC handlers that deserialize the I/O RPC calls' arguments, execute the I/O commands, serialize the results and return them to the client through the Libiofwd layer.

1.2.1 Network Stack

Currently two different network abstractions can be chosen for IOFWD: GASNet [2] and BMI [3]. Both offer methods to exchange large buffers, that ease programming of I/O RPCs and are user-space implementations. GASNet supports asynchronous communication and allows the registration of callbacks that are invoked when a buffer has been received. BMI doesn't offer callbacks and this feature had to be added. Both network abstractions offer support for several interconnects like Infiniband, Myrinet, MPI, TCP/IP and Portals. This allows some experimentation with communication networks on Linux-to-Linux I/O forwarding. On the SX-9 we were limited to TCP and UDP over 10 Gigabit Ethernet, the only network that can be used to connect an SX-9 node to Linux PCs.

The SX-9 uses a big-endian CPU, while Linux on x86_64 only works with little endian data. Therefore the GASNet UDP conduit (network driver) has been extended to support on-the-fly endianness conversion for the handshake protocol and header data, while treating transferred buffers opaquely. BMI supports connections between peers of different endianness as well.

1.2.2 Libiofwd

Libiofwd is a RPC library designed and developed within the project. It's API is independent from the underlying network abstraction and the most important calls are:

- iofwd_put_cmd() initiates a RPC call, i.e. transfers a small command buffer to the peer node and returns immediately.
- iofwd_wait() waits for the completion of an RPC call and/or drive network communication.

- iofwd_get_data() transfers the contents of a remote buffer to local memory synchronously.
- iofwd_put_data() transfers a local buffer's content to a remote node's memory synchronously.

The Libiofwd layer is endianness aware, it needs to "know" the endianness of the communication peers and convert commands and arguments appropriately.

1.2.3 Serializer + Deserializer

The Serializer and Deserializer libraries are based on the ZOID daemon from the ZeptoOS project [4] and were adapted for IOFWD. They are used to pack and unpack parameters and results of I/O operations to the format in which they are sent over the network with the help of Libiofwd RPCs to perform the remote I/O operation. The commands have been modified to deal with peers of different endianness.

1.2.4 Syscall Interception

Under Linux the relevant I/O library calls of an application could be replaced dynamically by preloading the syscall interception layer of IOFWD. The SX-9 does not support dynamic linking because it is a real memory system, like all vector machines. Instead of dynamic linking we have used the SX GNU linker ported within this project and its feature to wrap certain functions by renaming the original function name (e.g. *write()* to *__wrap_write()*). This way the libc and a part of the Fortran runtime libraries for the SX-9 have been relinked and I/O related functions were replaced by IOFWD wrapper functions. The IOFWD wrapper functions decide whether I/O operations will be executed locally, on the SX-9, or will be forwarded to the I/O forwarding server. The selection basis for local versus remote I/O is primarily based on the file path and is configurable, e.g. paths starting with */fwd* are handled remotely.

1.3 Implementation Status

IOFWD has been implemented successfully on top of two network abstractions: GASNet and BMI. The source code is available on the HLRS gforge site [1] and the IOFWD framework is installed on the SX-9 front-end machines as well as on the I/O forwarding servers accessible by all users at HLRS. IOFWD can forward around 60 I/O related library functions. It is easy to use and it supports C/C++ and Fortran applications with and without MPI.

1.4 Performance Results

The NEC SX-9 machines and the I/O forwarding servers communicate over a 10 Gigabit Ethernet network which is dedicatedly used for IOFWD.

As a starting point we compare the bare network stack performance between a SX-9 machine and an I/O forwarding server over 10 GE. GASNet uses UDP and reaches a bandwidth of 35 MB/s for send and 37 MB/s for receive with 4 MB I/O size on a SX-9 machine. BMI uses TCP and reaches 480 MB/s for send and 150 MB/s for receive with 4 MB I/O size on a SX-9 machine. The low GASNet bandwidth on the SX-9 is explained by the fact that the SX-9 operating system can't make use of the offload engine of the 10 GE card when communicating over the UDP transport protocol. In addition, the relatively slow scalar CPU of the SX-9 and high cost of context switches (interrupts, syscalls) due to the huge context that needs to be stored away (like vector registers) contribute their part to the very poor UDP performance.

BMI shows decent performance as it is able to use the 10 GE card's offload engine with the TCP protocol and therefore it is the preferred network stack for IOFWD on the SX-9.

More interesting from application point of view is the comparison of the performance of IOFWD with BMI to a Lustre filesystem with native Lustre performance on the I/O forwarding server and with Global Filesystem (NEC GFS) performance on the SX-9. The later filesystem is a SAN based parallel filesystem.

A Lustre client on the Linux I/O forwarding server at HLRS shows 500 MB/s write and 550 MB/s read performance. With IOFWD we achieve 280 MB/s write (as shown in Fig. 2) and 140 MB/s read (as shown in Fig. 3) performance into the same Lustre filesystem from a SX-9. Lustre is not natively supported by SuperUX, therefore IOFWD provides the most transparent way of accessing such a filesystem. The read performance is very close to the BMI maximum send transfer rate while the write performance suggests that there might be still room for optimization. The GFS performance has been measured just for comparison, on the HLRS setup a bandwidth of 500 MB/s and more for read/write is what an user can expect. However IOFWD cannot compete against GFS in terms of performance because it uses a different I/O network. It's main benefit is providing access to new filesystems for the SX-9 that are not supported natively by the SuperUX operating system.

2 IOFWD Usage

This section is an introduction of the usage of IOFWD, explains how an application can be modified to use it and shows some use cases and examples.

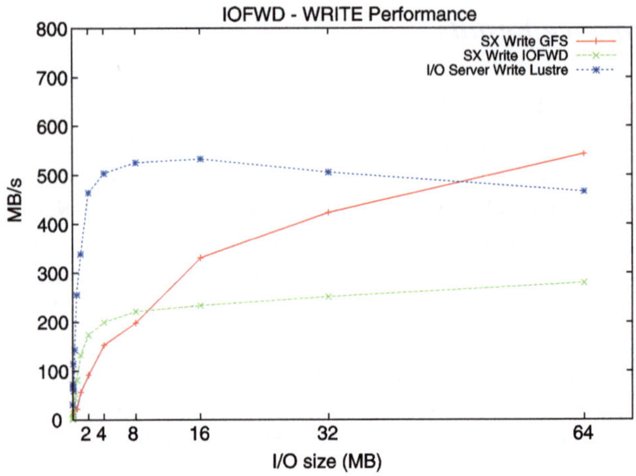

Fig. 2 IOFWD write performance on SX-9 to Lustre

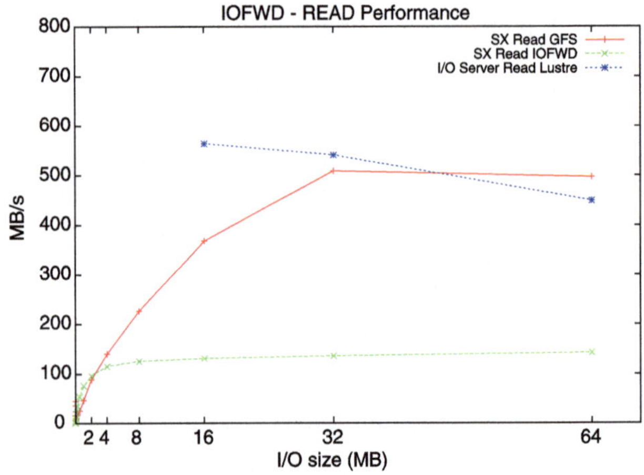

Fig. 3 IOFWD read performance on SX-9 to Lustre

2.1 System Overview at HLRS

The components involved in the IOFWD experiments at HLRS are sketched in Fig. 4:

- NEC SX-9: vector computing system for running simulations that are well vectorized,
- NEC TX-7: ia64 pre-processing and front-end nodes with large memory, used for accessing the SX-9 machine and pre-/postprocessing.
- Linux cluster used for post-processing and simulation runs of poorly vectorized applications,

- NEC GFS storage, holds the SAN based parallel filesystem mounted on the SX-9 and TX-7 nodes, connected to them through the fibre-channel fabric,
- Lustre storage, the primary filesystem of the Linux cluster, accessible from the TX-7 nodes through an NFS exporter,
- I/O forwarding servers that mount the Lustre filesystem and can run the iofwdsrv daemon.

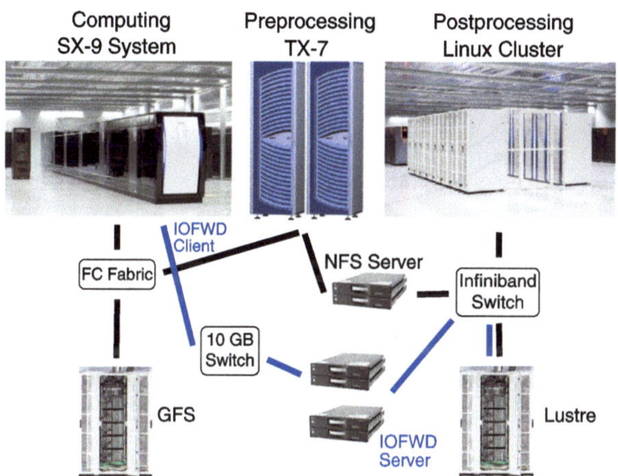

Fig. 4 System overview including IOFWD

2.2 Application Workflow Example

Figure 5 shows the normal data flow in an application from the user, pre-processing the input data on the TX-7, doing the simulation run on the SX-9 and finally running a post-processing step on the Linux cluster.

1. User submits data for preprocessing to the TX-7.
2. Data is being preprocessed, the output is written into the Lustre filesystem.
3. The pre-processed data represents the input data for the SX-9 simulation run, is copied from Lustre file system to the SX-9's GFS.
4. Pre-processed data from the GSF filesystem is used by the SX-9, the output of the simulation run is copied back to the GFS storage.
5. The simulation output is copied from the GFS storage to the Linux filesystem.
6. Simulation output is post-processed on the Linux cluster, the results stored back to the Lustre filesystem.

The copying steps (3 and 5) can spend a considerable amount of time and usually need to be done as part of the corresponding batch job, i.e. the simulation job or the post-processing job. This implies that the nodes reserved for the batch job are idle during the copying time. The longer the data files are, the longer the copying,

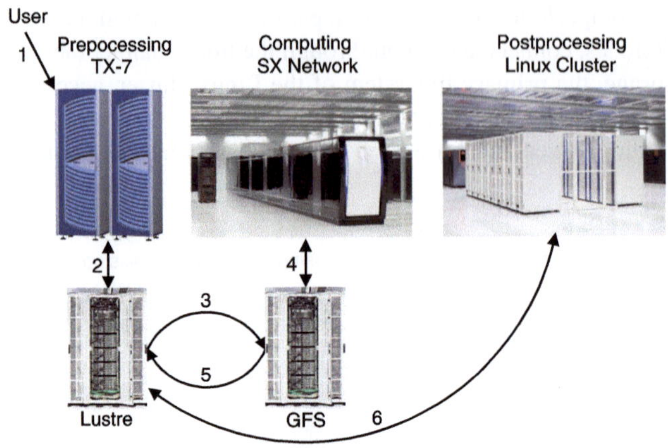

Fig. 5 Workflow of Application without IOFWD

thus the bigger the waste of energy and compute cycles. With I/O forwarding these copying steps can be avoided.

2.3 Application Integration, Compilation, Building and Running

Currently an application needs to initialize the IOFWD stack by explicitly calling a function, thus a small modification of the application's source code is required. Some place after the main program start and before doing I/O one should add the line:

```
call init_fwdadapt()
```

for Fortran programs, correspondingly

```
__init_fwdadapt(&argc, &argv)
```

for C/C++. In future we plan to do this initialization step at the first call of a forwarded I/O function, thus hiding it from the user and removing the need for an explicit initialization.

In MPI programs the initialization step has to be executed before the `MPI_Init()` call.

As mentioned in Sect. 1.2.4 the decision whether access to a file will be local (through the SuperUX operating system) or remote (through IOFWD) is influenced by the absolute path. The default path prefix for forwarded access is `/fwd`, this path prefix is currently compiled into the IOFWD server, but could be made configurable in future. It means that an access on the SX-9 to the file `/fwd/lustre/test` would be done remotely on the IOFWD server to the file `/lustre/test`. File paths used in applications need to be changed accordingly if access to remote files is desired.

The final step is using the IOFWD compile wrapper to compile and link the application that will use I/O forwarding. This hides the complexity of selecting I/O forwarding libraries. For C programs the compile wrapper call on the frontend node is:

```
IOFWD_CC=sxcc iofwd_sxc++ test.c -o test
```

while for Fortran one has to use:

```
IOFWD_F90=sxf90 iofwd_sxf90 test.f -o test
```

For MPI programs the compiler wrapper calls are:

```
IOFWD_F90=sxmpif90 iofwd_sxf90 test.f -o test
```

Applications using I/O forwarding are started like any other application. The system environment should set the environment variable `$IOFWD_SERVER` to the I/O forwarding server name or IP address. To improve scalability this variable can point to multiple servers, as explained in Sect. 1.1.

Unified Parallel C (UPC) is supported via the GASNet branch of IOFWD but isn't very well tested at the moment. For UPC applications it is possible to input the number of IOFWD servers via the environment variable `IO_SERVER_COUNT`. The addresses of the IOFWD servers and the compute nodes are given in a colon separated list in an environment variable called `SSH_SERVERS`. This list must first contain the addresses of the compute nodes, followed by the addresses of the IOFWD servers. The IOFWD clients are automatically assigned in round robin manner to the IOFWD servers during the initialization of IOFWD. The initialization routine for IOFWD is the same as for C. A simple wrapper exists to compile an UPC application for IOFWD.

2.4 First Real Application Experiences

IOFWD was tested with two real applications namely CPMD [6] and KOP3D [7].

In CPMD a short simulation sequence is followed by the writing of a restart file via BMI IOFWD. The restart file can be used for analysis/post-processing or for restarting the simulation from the point it represents. The example we ran writes a restart file of 7.5 GB. The elapsed run time on SX-9 with writing to the fast local GFS storage was 825 s that included 20 s of I/O time. When using I/O forwarding over BMI the elapsed time grew only minimally to 837 s, out of which 30 s were spent for I/O. The increase in I/O time was expected and is explainable by the slower performance of the remote writing compared to the local fast GFS access. The minimal increase in I/O time (10 s) is still a very good result as it makes the transfer of the data to the post-processing system unnecessary, a transfer that would take an order of magnitude of 20–30 s, during which typically all nodes reserved for post-processing would be idle.

The KOP3D application is a coupled simulation that embeds a structured grid into an unstructured grid (or vice-versa) and can run the different simulations on different machines. In our case the structured part is well vectorizing and running on the SX-9 vector system while the unstructured part runs on the Linux cluster. I/O forwarding can be used to simplify the workflow of such a complex application by keeping and creating all files on one single filesystem. In our example KOP3D used a startup file of about 60 MB that it loaded through IOFWD on the SX-9 side. The performance impact of IOFWD was not noticeable, and KOP3D proved to be a hard testcase for IOFWD that helped revealing and fixing a few bugs.

3 Conclusion

We have designed and implemented an I/O forwarding framework that allows programs on the SX-9 to use filesystems and storage that are neither supported by the operating system nor directly connected to the machine. The main benefit of the I/O forwarding approach is the offloading of the I/O activity to nodes that are better suited for that, thus optimizing the use of cycles and power on the SX-9 vector machine, which is highly specialized for computational tasks. The offloading of I/O to Linux nodes allows users to optimize their workflows and administrators to consolidate the various different filesystems available in their datacenters. For SX-9 installations it enables the use of less expensive storage which leads to the reduction of the demand for the expensive fibre-channel SAN based parallel filesystem.

As an outlook, the project will continue to optimize the I/O forwarding approach and work towards simplifying the use of IOFWD and extending its scope to other machines than SX-9. I/O forwarding is a concept that can be applied to any setup where compute nodes are highly specialized, for example in very large Linux clusters deploying simplified operating systems on the compute nodes. The concept can be applied as well to applications running on coprocessors.

Furthermore we plan to develop I/O analysis tools that help moving applications to IOFWD and analyse the library and system calls invoked. An extension of the concept allowing for on-the-fly file content manipulation and a more generic toolchain for applications coupling is envisioned, too.

The IOFWD framework for SX-9 can be downloaded at [1].

References

1. HLRS GForge, https://gforge.hlrs.de/projects/sxlinux/
2. Global-Address Space Networking (GASNet), http://gasnet.cs.berkeley.edu
3. Buffered Message Interface (BMI), http://www.pvfs.org
4. ZeptoOS, http://www.mcs.anl.gov/zeptoos/
5. SX Linux, http://code.google.com/p/sx-gcc/
6. CPMD, http://www.cpmd.org/
7. Heterogeneous Parallel Aero-Acoustics Using PACX-MPI, http://www.springerlink.com/content/u46125476w324380/

High-Speed Data Transmission Technology for the NEC SX-9

Hiroshi Yamaguchi, Hiroshi Takahara

Abstract The SX Series has an advanced architecture of large-scale shared memory, high-speed data transfer between the CPU and the memory, and an ultra high speed network interconnecting nodes. One of its key technologies in achieving high system efficiency is the fast data transmission between LSIs making up the system, which highly relies on the LSI and circuit technologies, as well as their inspection technologies. NEC has developed sophisticated technologies for CMOS, high-speed interface for efficient data transfer, high speed and low skew clock distribution, and noise reduction. This paper outlines key technologies that enable high performance of the SX-9 supercomputer on real scientific applications from the view point of fundamental technologies.

1 Introduction

The SX Series is the parallel vector supercomputer that has continuously been enhanced by NEC toward the SX-9, which features the world fastest CPU core of 102.4 GFLOPS, the large-scale shared memory up to 1 TBytes, and the interconnects with a data transfer rate of 128 GBytes/sec. It makes up one node system of a 1.6 TFLOPS peak performance, which can be configured up to 512 nodes with the maximum vector performance of 839 TFLOPS [1, 2]. In addition, the improved LSI technology and high-density packaging technology have reduced both power consumption and required installation space to approximately a quarter of that required for conventional supercomputers.

Hiroshi Yamaguchi
IT Hardware Operations Unit, NEC Corporation, 1-10 Nisshin-cho, Fuchu, Tokyo 183-8501, Japan
e-mail: h-yamaguchi@cw.jp.nec.com

Hiroshi Takahara
HPC Division, NEC Corporation, 1-10 Nisshin-cho, Fuchu, Tokyo 183-8501, Japan
e-mail: h-takahara@bc.jp.nec.com

M. Resch et al. (eds.), *High Performance Computing on Vector Systems 2010*,
DOI 10.1007/978-3-642-11851-7_5,

In order to implement even more impressive performance with the SX-9, NEC has advanced the LSI and circuit technologies. Since an increase in the inter-LSI signal transmission speed is critical for improving the system performance, NEC has also developed a new multichannel serial interface for the high-speed inter-LSI data transfer on the SX-9. Moreover, both the power consumption and the amount of interface circuitry have been reduced in order to enable the mounting of multiple channels on an LSI [3].

The improvement of the processing capability of a high-speed system requires an increase in the speed both of the intra-LSI and the inter-LSI signal transmissions. The countermeasures for the power noise that hinders increases in signal speed are also an important factor. In order to transmit signals stably at high speeds, the SX-9 uses wiring boards made of low-loss materials with low signal attenuation during transmission and some of its circuits incorporate an equalization function to improve the transmission signal waveforms. The SX-9 reduces power noise by optimizing the number of decoupling capacitors with regard to the increase in power noise that occurs as a result of the increased frequency of changes in supply current due to the use of higher-speed transistors.

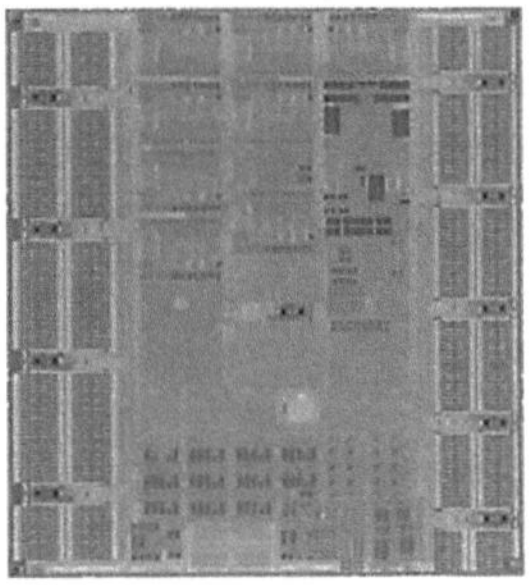

Fig. 1 External view of the SX-9 CPU chip

Table 1 Specifications of the SX-9 CPU chip

Item	Specification
Process technology	65 nm
Number of transistors	350 million
Supply voltage	1.0 V
Number of pins (signal pins)	8,960 (1,791)
Wiring layer configuration	11 copper layers
I/O interface	CML
Packaging	Bare chip configuration

2 LSI Technology

The external view and the specifications of the CPU chip of the SX-9 are shown in Fig. 1 and on Table 1, respectively. The common specifications of the LSIs used with the SX-9 include the 65 nm CMOS process, 11-layer copper wiring, improvements in routing delays by adopting inter-layer dielectric films, etc., implementation of large-capacity on-chip capacitors by developing the MIM (Metal-Insulator-Metal) process, and implementation of a high-performance low-voltage supply by reducing the thickness of the gate oxide film. Low transfer latency is also achieved by developing a new multichannel serial interface.

2.1 Serial Interface

The SerDes (Serialization/De-serialization) technology with a transmission rate of 10 Gbps has been developed. Here an electric transmission technology is employed for connecting multiple channels between LSIs. As the signal is attenuated to about one-tenth of its original intensity due to the waveform distortion produced by the frequency characteristics of the PWB (Printed Wiring Board) and the cables in the transmission paths, a pre-emphasis (EMP) circuit in the transmitter (TX) and an equalizer circuit in the receiver (RX) are incorporated. In addition, a sampling oscilloscope (iSCOPE) function is incorporated into RX in order to enable the eye diagram sampling of the receivable voltage and timing, while RX is mounted in the system.

The input voltage is compared to the reference voltage by using the iSCOPE function. Based on the comparison of the voltage with those just before and after the current one, the computed EMP control outputs are then sent to the diagnostic processor. These signals are used, when the system is initialized in order to create the EMP control signal reflecting the effects of the data before and after the pre-emphasis of TX and the iSCOPE functions reference control signal, which are then sent to the SerDes circuit. The eye can be maximized by repeating this adjustment. TX and RX occupy a small area of 0.31 m^2 per channel, and about 400 TX/RX channels are accommodated per LSI.

2.2 Clocks

The distributed clocks can roughly be divided into the clocks for logical circuitry and those for interface circuitry. The logical circuit clocks are distributed with the 2-step method. The clocks distributed over a wide area are wired using the clock-dedicated thick film wiring layer with low resistance in order to reduce the waveform skew due to low resistance, and are distributed with equal delays based on considerations for the inductance component of the wiring as well as for its RC component.

The clocks distributed in a narrow area are distributed with equal delays and the clock skew is reduced by optimizing the driving power of the clock drivers. In addition, to reduce the power consumption, a clock gate circuit that multiplies part of the clock as required is adopted.

The interface clocks use the CML (Current Mode Logic) circuitry. The CML circuitry uses differential signals and drives at a constant current, so that it can increase the clock frequencies and greatly reduce clock jitter due to power noise compared to the CMOS inverter circuit.

In order to generate high-speed clocks, the SX-9 adopts the APLL (Analog Phase-Locked Loop) circuit that multiplies the clock input from outside the LSI. The APLL incorporates LC tank type VCO (Voltage-Controlled Oscillators) composed of inductors and variable capacitances that are fabricated on the LSI and generates a clock obtained by aligning the phases of the external and internal clocks of the LSI. For countermeasures against power noise, a power regulator is built in to reduce jitter by generating the LSI-dedicated independent power supply inside the LSI.

3 High-Speed Circuit Technology

Improvement of the processing capabilities of a high-speed system necessitates an increase in the inter-LSI signal transmission speed, as well as in the intra-LSI signal transmission speed. The countermeasures against power noise that hinders increases in the signal transmission speed are also important. There are five key technologies for high-speed data transmission as indicated below.

1. Waveform shaping
   Measures against signal attenuation/waveform distortion
2. On-chip observation
   Observation of incoming waveform, power-supply noise, temperature, and signal deterioration within LSI
3. Elimination of skew
   Skew-free signal propagation by clock regeneration
4. Low-power design
   Reduction of serialized paths
   Optimum selection and design of circuits
5. Noise suppression
   Noise reduction with differential transmission
   Design with isolated power supply

In this section, items 1, 2, and 5 above are described.

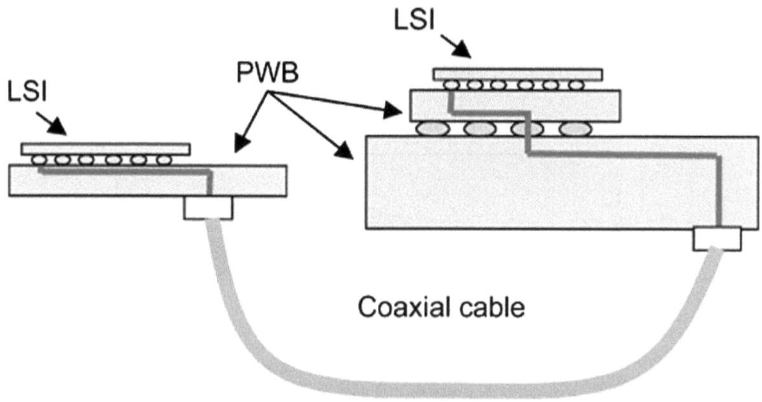

Fig. 2 Connection configuration between LSIs of the SX-9

3.1 Transmission Technology

The inter-LSI signal transmission attenuates the high-frequency component of a signal due to the skin effect and the dielectric loss. This makes correct signal reception impossible, when the wiring length is large.

As shown in Fig. 2, the structure of the inter-LSI connections of the SX-9 tends to increase the attenuation because of the connections between the multiple wiring boards. A pre-emphasis function has been added to ensure correct signal reception in such a transmission path structure. The pre-emphasis function has been enhanced for the SX-9 by increasing the number of amplitude adjustment steps and by introducing an algorithm for the automatic setting of the optimum step value according to the results of the monitoring of received waveform.

In addition, the wiring boards are made of low-loss materials with low signal attenuation and low-loss cables for the connection between the wiring boards that have been developed to reduce the overall attenuation via the transmission path.

In determining the structure of a transmission channel, the simulation was made for the transmission path along with the evaluation of the LSIs for use in testing. With the transmission path simulation, a system for modeling the three-dimensional structure of the wiring board was introduced based on the electromagnetic field analysis in order to improve accuracy especially for high frequencies. The electromagnetic field simulator utilized here allows us to conduct evaluations for various conditions of the wiring board, such as wiring length, via hole, and pad shape. In this way, it is made possible to implement a transmission path structure capable of offering an eye aperture of the reception waveform that satisfies the design target.

The simulated receiving waveform for transmission path was compared with the waveform observed for the transmission via the actual transmission channel, which verifies the high accuracy and effectiveness of such simulations (Fig. 3). The inter-

LSI transmission with the path structure as shown in Fig. 2 has been implemented based on the investigations as outlined above.

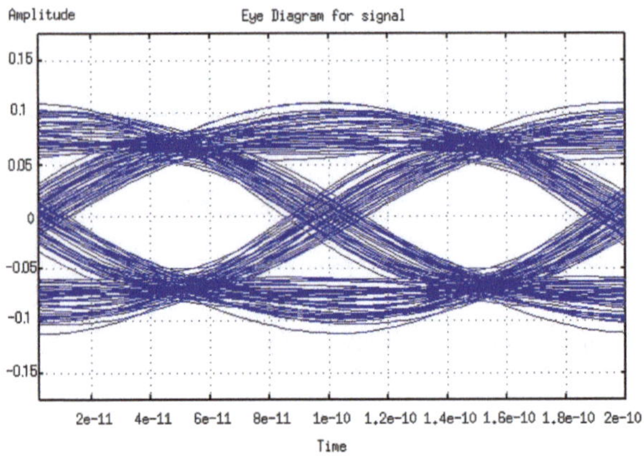

Simulated waveform

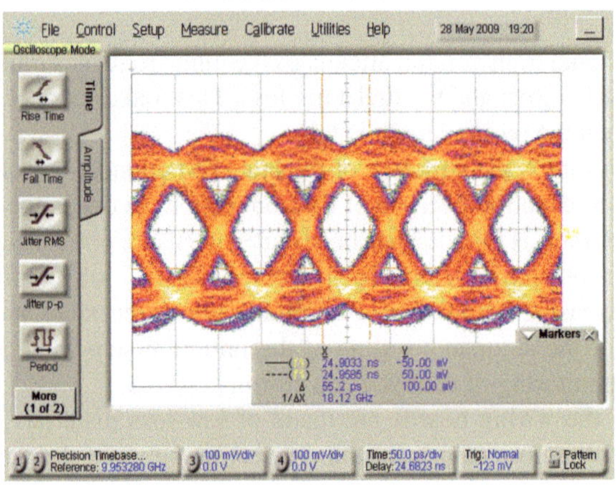

Measured waveform

Fig. 3 Simulated and measured waveforms

Figure 4 illustrates waveform shaping based on the pre-emphasis technology. While the eye pattern can not be obtained without wave shaping, optimal eye patterns can be obtained by exploiting the pre-emphasis. This is an essential technology used for high-speed transmissions at a Gbps level.

Figure 5 shows on-chip waveform observation. Here eye patterns can be seen with the sampling oscilloscope. Any expensive tester is not required because of the LSI circuit-mountable apparatus, enabling the observation of real waveforms.

The actual input waveform within the LSI is not visible, because there is decay in the LSI. It can only be observed by using the on-chip sampling oscilloscope (iSCOPE).

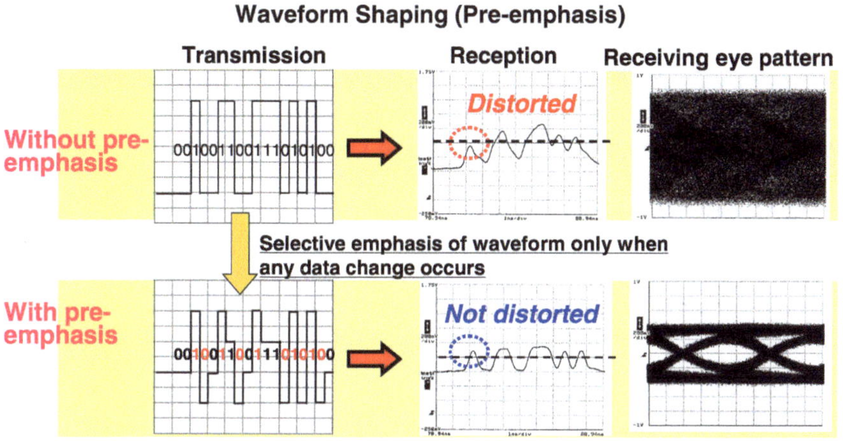

Fig. 4 Waveform shaping with the pre-emphasis

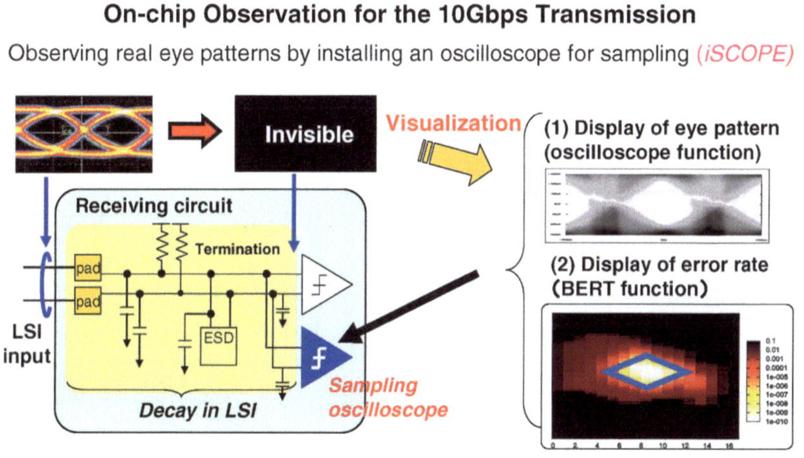

Fig. 5 On-chip waveform observation for the data transmission at 10 Gbps

3.2 Power Noise Countermeasures

In order to achieve ultra-high-speed transmission, it is necessary to adequately reduce power noise, decrease the jitter superimposed on the transmitted waveform, and ensure the margin of the circuit operation. With the SX-9, quantitative investigations were conducted for the propagation of the power noise generated in the core logic circuit to the ultra-high-speed I/O by carrying out an entire configuration analysis from the wiring board to the LSIs. As a result, the value of on-chip capacitance mounted in the LSIs can be determined so that the propagation does not affect the SerDes circuits.

While a sufficient capacitance value can not be assured for the SX-9 with the capacitance created using the gate, the MIM capacitance can be placed on the highest layer of the LSI so that the MIM capacitance is uniformly distributed all over the chip surface for the efficient and effective noise reduction.

The actual effects of the noise reduction are examined by using the circuits implemented within the LSI for observing noise and jitter and by applying the specified amount of jitter using a clock modulator, thus enabling the reservation of a margin in the inspection.

On the basis of the electro-magnetic simulation, it is possible to determine as to where the capacitors need to be placed (Fig. 6). The temporal change of the current for the entire chip is calculated as a current waveform on the basis of the timing data of signals moving through flip-flop circuits and gates. Then, the physical structures of LSI and PWB can be modeled by using the CAD data for the LSI and the board.

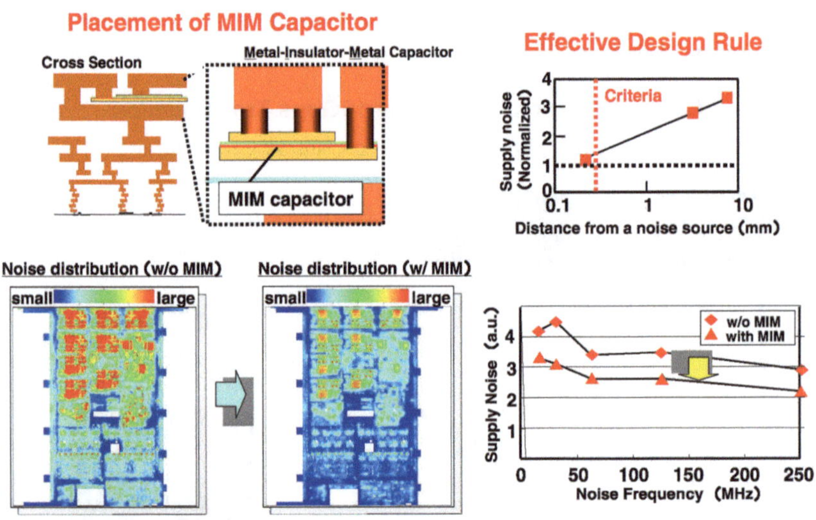

Fig. 6 Evaluation of the MIM capacitance

As indicated at the upper right, the capacitor needs to be placed close to the noise source for reducing the power supply noise. On the other hand, in many cases it is difficult to place such a capacitor at an ideal place, since there is a logic circuit with the ordinary gate capacitor. The MIM technology allows such capacitors to be placed appropriately within the LSI. As a result, the supply noise can successfully be reduced by 25noise over a wide range of frequency bands. The upper left depicts the MIM with the uppermost metal layer zoomed. The MIM is formed between the uppermost layer and the layer underneath it.

The measurement of power supply noise is made by using an on-die detector placed within the LSI, as indicated in Fig. 7. The upper left represents the schematic view and its operation is at the bottom left. After applying a noise, a judgment is made as to whether the power supply noise level is greater than VREF (reference voltage) or not by utilizing the comparator.

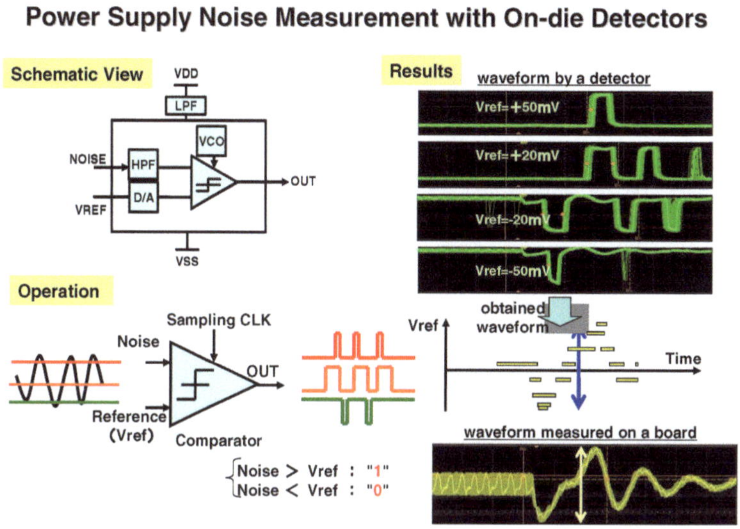

Fig. 7 Measurement of power supply noise

Actually, the waveform can be obtained for each VREF, as shown at the upper right. By superimposing these results in a pattern as represented at the center, it turned out that such a pattern agrees with the measurement below. In such a way, the model can be verified through the combination of simulation and measurement.

4 Summary

The outline of the LSI, circuit, and inspection technologies for the SX-9 has been given. As indicated in this paper, the high-speed signal transmission is key to the high performance of the SX-9 supercomputer. NEC continues to enhance such technologies for future advanced systems.

References

1. http://www.nec.de/products-services/high-performance-computing/index.html NEC SX-9 Supercomputer (2008)
2. Takahara H., HPC architecture from application perspectives. In: High Performance Computing on Vector Systems 2009 (edited by Resch, M. et al.), Springer, 59–67 (2009)
3. Tanahashi T., et al., LSI and circuit technologies of the SX-9. NEC Technical Journal (Supercomputer SX-9/Special Issue), Vol. 3(4) http://www.nec.co.jp/techrep/en/journal/g08/n04/080405.html (2008)

Part III
Grid and Cloud Computing

The Vector Computing Cloud: Toward a Vector Meta-Computing Environment

Ryusuke Egawa, Manabu Higashida, Yoshitomo Murata, Hiroaki Kobayashi

Abstract Aiming at realizing a high performance computing (HPC) cloud with vector supercomputers, this paper presents the world's first prototype of wide-area vector meta computing infrastructure named a **vector computing cloud** by virtualizing remote vector computing resources as an HPC service over the Internet. The prototype system consists of two remote NEC SX-9 nodes connected through a long fat-pipe network (LFN), and each node is located at Tohoku University and Osaka University with a distance of 800 km. The vector computing cloud also provides a single sign on environment and jobs are automatically assigned to appropriate sites. Wide-area co-operation of distributed vector supercomputers is realized by adopting the NAREGI Grid Middleware, and a virtual machine for NEC SX vector supercomputer series, job scheduling algorithms, and an MPI operating environment are newly developed to enhance the job and resource management capabilities of the NAREGI Grid Middleware. In addition, to achieve fairness and efficient job scheduling on the vector computing cloud, this paper presents a history-based job scheduling mechanism for a queue system. Based on the estimation, the job scheduling mechanism automatically allocates the job to an appropriate site, which can execute the job earlier. The operation tests and experiment results indicate that the prototype system realizes single sign on for multiple vector resources and has enough potential for transparently operating jobs between the two SX-9 systems. This paper also evaluates and discusses the performance of the proposed job scheduling mechanism and MPI operation between both SX-9 systems using the High Performance Linpack (HPL) benchmark.

Ryusuke Egawa, Yoshitomo Murata, Hiroaki Kobayashi
Cybercience Center, Tohoku University, 6-3 Aranaki Aza Aoba, Sendai, Japan, 980-8578
e-mail: egawa@isc.tohoku.ac.jp, murata@isc.tohoku.ac.jp, koba@isc.tohoku.ac.jp

Manabu Higashida
Cybermedia Center, Osaka University, 5-1 Mihogaoka, Ibaraki, Osaka, Japan, 567-0047
e-mail: manabu@cmc.osaka.ac.jp

M. Resch et al. (eds.), *High Performance Computing on Vector Systems 2010*, 75
DOI 10.1007/978-3-642-11851-7_6, © Springer-Verlag Berlin Heidelberg 2010

1 Introduction

The current high performance computing (HPC) applications require ever-increasing computational power. To execute such an HPC application on existing computing resources, one way can conceive to make co-operation among distributed supercomputers by GRID technologies. However, a grid environment still forces HPC users to consider available computing resources to running huge and massively parallel HPC applications. In the conventional distributed HPC systems, users should select an appropriate site to execute their codes and submit HPC jobs directly by themselves [1].

Recently, the cloud computing attracts a great deal of interest as a new generation IT infrastructure. Although there are many definitions of the cloud computing [2, 3], the cloud computing is the internet-based computing in the broad sense. In the cloud computing, the resources, infrastructures and software are provided as a cloud service through the internet. Here, these cloud services are virtualized, and the cloud users need not attend to the actual resources and infrastructures. So, an HPC service provided by the cloud computing, which is called a HPC cloud, is seen as a high possibility to execute massive parallel applications [4, 5]. In the HPC cloud, users would not have to care where their jobs should be operated since the supercomputers are virtualized as a huge single system and jobs are automatically assigned to appropriate sites.

Vector supercomputer systems distributed over the world have been operated with high utilization ratio due to its high-sustained performance. Especially, the vector supercomputer demonstrates its high potential in large scale scientific computing such as fluid dynamics, structural dynamics, climate simulations, and so on [6, 7]. However, more convenient and powerful vector computing environments are needed in order to boost advanced scientific researches. Under this situation, a vector meta-computing infrastructure is expected to realize efficient usage of computing resources and to run more massively parallel simulations across multiple vector supercomputing systems. In addition, from user's viewpoints, the infrastructure should keep the easiness to use computing resources as recent cloud systems [8, 9].

Aiming at realizing an HPC cloud with vector supercomputers, this paper presents the world's first prototype of wide-area vector meta computing infrastructure named a vector computing cloud by virtualizing remote vector computing resources as an HPC service over the internet. The prototype system consists of two remote NEC SX-9 nodes connected through a long fat-pipe network (LFN), and each node is located at Tohoku University and Osaka University with a 800 km distance. The vector computing cloud also provides a single sign on environment and submitted jobs to the vector computing cloud are automatically assigned to appropriate sites. Wide-area co-operation of distributed vector supercomputers is realized by the NAREGI Grid Middleware, and a virtual machine for NEC SX vector supercomputer series, a job scheduling mechanism, and an MPI operating environment are newly developed to enhance the job and resource management capabilities of the NAREGI Grid Middleware. The organization of this paper is as follows: Sect. 2 provides background and the basic concept of the vector computing cloud. Section 3 describes the prototype system of the vector computing cloud newly developed GridVM for SX and a job scheduling mechanism for the vector computing cloud. In Sect. 4, the performance and potential of the prototype system are discussed. Section 5 concludes this work.

2 Basic Concept of the Vector Computing Clouds

Though there are strong demands for the HPC cloud to execute huge scale massive parallel applications from an HPC community, there are few challenges to blend HPC and clouds [8]. That is why, it is simply would not be a profitable business due the niche market and technical obstacles to treat complex workflows of HPC applications. However, from the viewpoint from academia, the HPC cloud infrastructure that can process HPC applications is needed to boost scientific and engineering innovations. This paper focuses on the high-sustained performance of vector supercomputers and tries to develop the HPC cloud with vector supercomputers named the *vector computing cloud*.

To satisfy the HPC users requirement, the vector computing cloud should have characteristics as follows;

1. "*single-sign-on*" and "*ease-to-use*" environments via internet
2. smart job scheduling which can effectively use distributed computing resources, though they are operated by different job execution policies
3. capability of handling large-scale applications that requires high computing power which surpasses potential of each supercomputing system that constructs the vector computing cloud.

The distributed vector computing resources are co-operated by grid technologies with newly added portal interfaces. Thus, the vector computing cloud realizes a single sign on for multiple vector computing systems by virtualizing vector computing resources. In conventional methods, users should select an appropriate site to execute their codes by themselves as shown in Fig. 1. On the other hand, in the vector computing cloud shown in Fig. 2, users do not have to care where their jobs should be operated since the vector supercomputers are virtualized as a huge single system and jobs are automatically assigned to appropriate sites. To realize efficient scheduling among vector supercomputers, an enhanced job scheduling mechanism is introduced in this work. In addition, the vector computing cloud is expected to not only improve system dependability by making multiple systems work in a complement style, but also realize a large-scale simulation that cannot execute on individual vector supercomputers that compose the vector computing cloud. To operate such a job, the vector computing cloud provides a GridMPI execution environment [10].

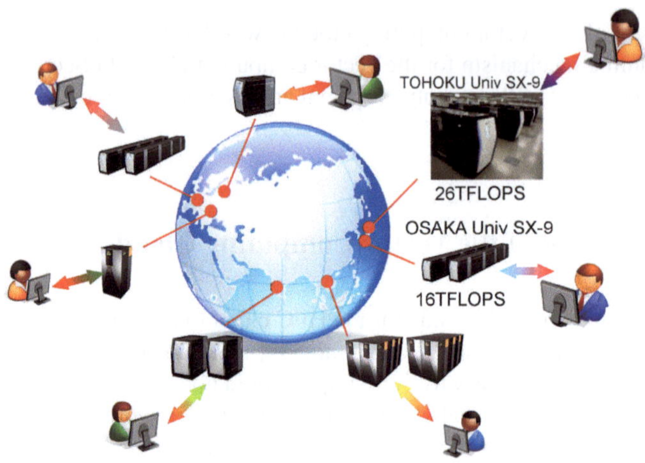

Fig. 1 Distributed vector computer systems

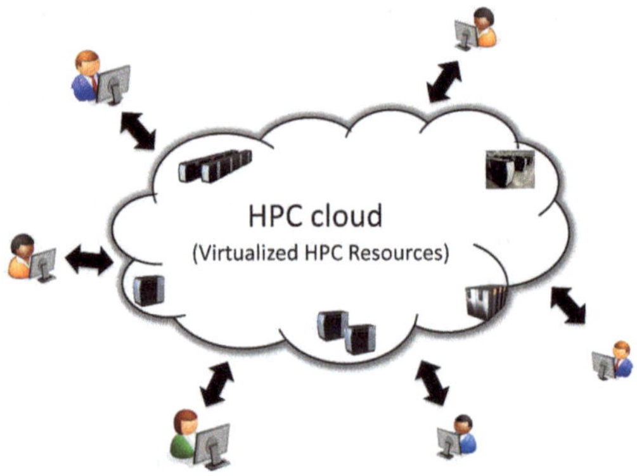

Fig. 2 The vector computing cloud

3 Prototyping of the Vector Computing Cloud

The vector computing cloud is designed based on the NAREGI Grid Middleware [11], which enables large-scale mete-computing by gathering the geographically distributed computing resources. Since the NEC SX series have not been fully adapted to NAREGI Grid Middleware, a new component called GridVM for SX is developed. Not only for the adaption, GridVM for SX also enhances the capabilities of the NAREGI Grid Middleware in terms of job and resource managements.

As the first step in establishing the vector computing cloud, the prototype system is implemented as shown in Fig. 3. The prototype consists of two nodes of SX9 vector supercomputers. Each node is located at Tohoku University and Osaka University. Both nodes have 16 CPUs with 1 TB memory and are connected through a science information network (SINET3). SINET3 is an information com-

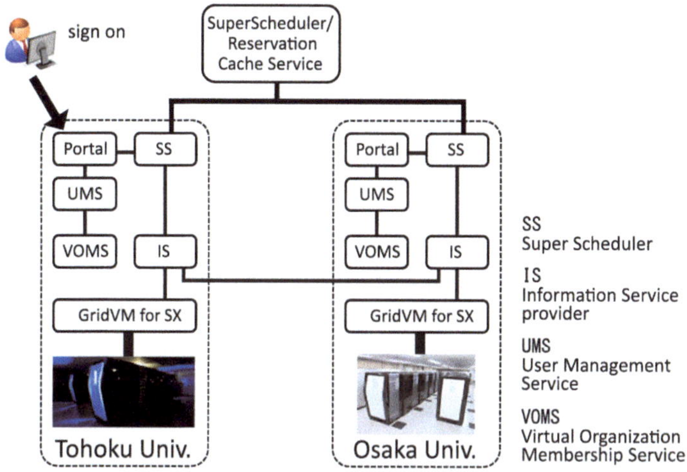

Fig. 3 System configuration of prototype system

munication network, which connects universities and academic research institutions in Japan [12].

The vector computing cloud consists of multiple sites with vector supercomputers, each of which has several components of NARGI Grid Middleware: Portal, User and Virtual Organization Membership Service (UMS/VOMS), Information Service (IS), Super Scheduler (SS), and GridVM for SX. Portal provides a web interface of the virtualized system, and UMS/VOMS authenticates the users and servers. IS manages the resource information of each site with gathering the utilization status. IS also communicates with IS's in other sites to share utilization statuses. SS searches computing resources for user requests, and allocates jobs based on the information from IS. Reservation Cache Service (RCS) performs as a global scheduler among each site. RCS aggregates and controls the requests from SS's to find out an appropriate site to execute a job. GridVM for SX is a virtual machine for SX vector supercomputers, which performs synchronization control of vector computing resources in the site, and provides mete-computing environments based on with high affinity with a local job scheduler. GridVM for SX keeps high compatibility with the local job scheduler NQS II [13] on the SX-9, which enables the efficient use of vector computing resources even in the vector computing cloud. Moreover, it permits the co-existence of conventional jobs and cloud jobs that running on the vector computing cloud.

3.1 Virtualizing Vector Supercomputers: GRID VM for SX

The GridVM of NAREGI Grid Middleware virtualizes the computing resources by interfacing with the local scheduler, which actually manages the computing resources, and supplies the computing resources to the grid environment. More detailed information of NAREGI Grid Middleware are well described in [14]. So far, GridVM have not adapted to Super/UX for SX vector supercomputers, we newly develop GridVM for SX with three additional features. First, GridVM for SX improves an information gathering function based on Ganglia [15] by employing direct communication between GRID VM for SX and IS. Second, considering practical usage of NAREGI on supercomputer systems, co-existence of reserved job by NAREGI and local job of supercomputers is permitted. To realize the co-existence, a job assignment map of a submitted job to a local scheduler NQSII is synchronized with a reservation map of SS. Third, allowing the execution of MPI/SX which is a dedicate MPI for NEC SX series.

3.2 Job Scheduling on the Vector Computing Cloud

In the vector computing cloud, each supercomputer has their own job execution policy. Then, the execution of a job submitted to the vector computing cloud also complies with the policy of allocated site. The difference in the policies makes job scheduling on the vector computing cloud complicated. In the prototype system, SX-9 nodes at Osaka University employ a reservation system for job executions, and the execution time of jobs is limited. This reservation-based operation guarantees the time when a job execution starts (*job-start time*). Thus, the job-start time can easily be obtained by checking a reservation map of the system. On the other hand, SX-9 nodes at Tohoku University employ a queuing system for job execution in a FIFO manner and the system does not limit the execution time of jobs. The queuing-based operation allows running large scale jobs with no time limitation and provides the high-utilization of computing resources without reserving and allocating excess resources for small-size jobs. However, this queuing-based system cannot guarantee job-start time, because the execution time of jobs in a queue is undetermined.

From the viewpoints of users, the vector computing cloud should execute a job as soon as possible from the time when user submitted a job. However, if a scheduler in the vector computing cloud that consists of computing resources with different job operating policies could not understand job-start times of both job execution systems, the scheduler can not allocate a submitted job to a computation resource, which can execute the job earlier. Therefore, to achieve such a job scheduling on the vector computing cloud, estimating the job-start time in the queuing-based system plays important role.

To overcome the job scheduling problem mentioned above, this paper proposes a job scheduling mechanism which estimates the job-start time in a queuing-based system and allocates a job to an appropriate site in the vector cloud computing environment. Figure 4 shows the overview of the proposed job scheduling mechanism. The job scheduler consists of original NAREGI modules; SS and Reservation Map, and newly added SS's sub-module named *resource select module*. The *resource se-*

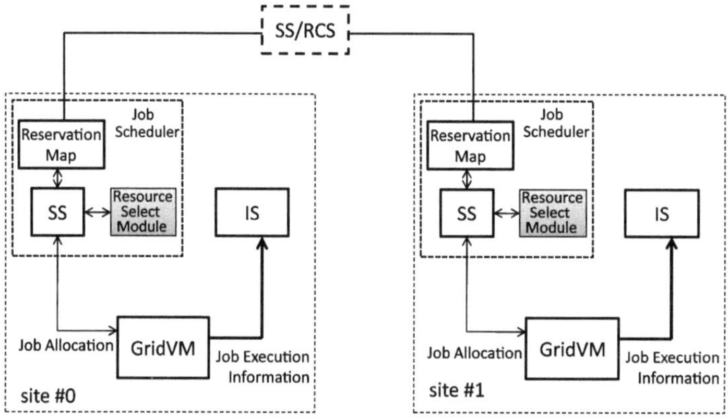

Fig. 4 Overview of job scheduler

lect module is invoked by SS, and estimates a job-start time from the history of job-execution on vector supercomputers.

3.2.1 Job-Execution Time Estimation for a Queuing-Based System

To estimate the job-start time for a queuing-based system, we focus on the high software reusability in HPC. For example, a parameter sweep experiment, which is one of the famous HPC jobs, executes one program with many different parameters. Then, it is easy to estimate the execution time of the program by using the previous execution result.

We use a *job-execution time* of a job in a queue for a job scheduling in the vector computing cloud. The job-execution time indicates a required period to process the job. If we can obtain the job-execution time, we can also estimate job-start time by summing up job-execution time in the queuing-based system. By comparing the job-start time in a queuing-based system and that in a reservation-based system, the scheduler can assign the job to an appropriate site, which can execute the job early.

The scheduler records and archives job execution information to estimate the job-start time of the following jobs. When a job execution is completed, Grid VM sends the job execution information to IS. IS stores the information in a database, and provides an database access interface to SS. Then, SS can obtain the job-execution time based on the job name, job type, and the user account information. The *resource select module* obtains the job-execution time in a queue by using the job information accumulated in the database of IS.

Figure 5 shows the process of estimating the job-execution time. The process to obtain the job-execution time in a queue by the scheduler is described as follow. First, the *resource select module* accesses Grid VM, and obtains a list of queued jobs. Note that these jobs are not executed, and the job-execution time has not been

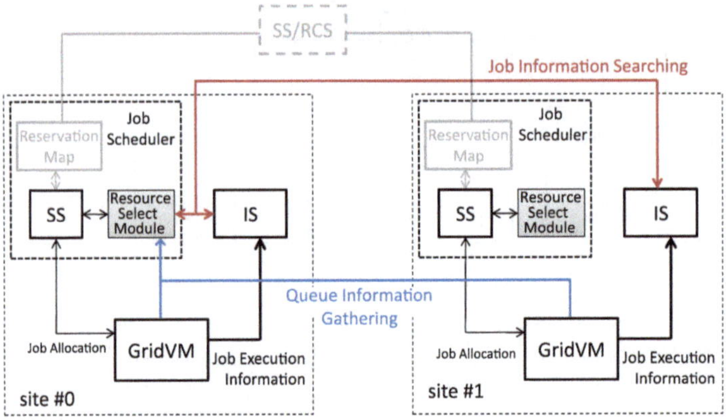

Fig. 5 Process of estimation

decided yet. Next, the *resource select module* retrieves the execution time of all jobs in the queue from IS. In this process, the *resource select module* uses the command name as the search key, and retrieves the execution time of the corresponding job which has the same command name. If the *resource select module* cannot obtain the job-execution time from the database of IS, the *resource select module* uses the average job execution time accumulated in IS. Finally, by adding all job-execution time in the queue, the *resource select module* estimates the job-start time for all jobs in the queuing-based system.

3.2.2 Job Allocation Mechanism on the Vector Computing Cloud

The job submitted by the user is allocated to a computation resource by SS. This subsection describes the processes of job allocation on the vector computing cloud.

First, SS obtains a list of computation resources, which satisfy the requirement of the job such as the number of processors, memory capacities and so on. Next, to select an appropriate resource from the list of computation resources, SS calls the *resource select module*.

Figure 6 shows a situation for selecting an appropriate one from the queuing-based site and the reservation-based site to allocate a new job. The *resource select module* obtains the job-start time of a queue in the queuing-based site and the reservation-based site. The job-start time of the reservation-based site can be obtained from the reservation map. Then, the *resource select module* compares the job-start time of the queuing-based site with that of reservation-based site, and allocates the new job to the computing resources which can execute the job earlier. In Fig. 6, the *resource select module* allocates the new job to the reservation-based site because the reservation-based site becomes ready to execute the new job earlier than the queuing-based site.

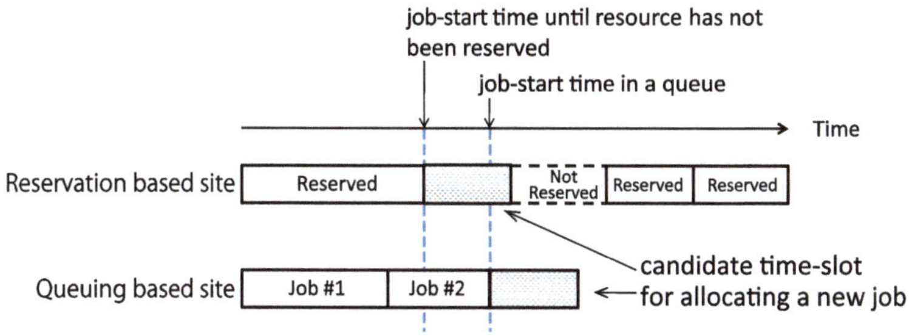

Fig. 6 Job allocation to a reservation-based site and a queuing-based site

3.3 MPI Environment for Vector Computing Cloud

As mentioned in Sect. 3.2, the vector computing cloud allows to operate MPI/SX as a local MPI and GridMPI as local and global MPI. Figure 7 shows an MPI execution environment of the vector computing cloud. In each site, the two vector supercomputer SX-9 are connected through a high speed cross bar switch named IXS and an 1 Gb Ethernet(GbE), and the both MPIs are executable. The MPI/SX is optimized for IXS to achieve superior performance in SX series. Furthermore, to achieve higher performance even in the GridMPI, a wrapping interface to MPI/SX is implemented. In addition, GridMPI requires an IMPI server to control MPI communications between the supercomputing nodes [10, 16].

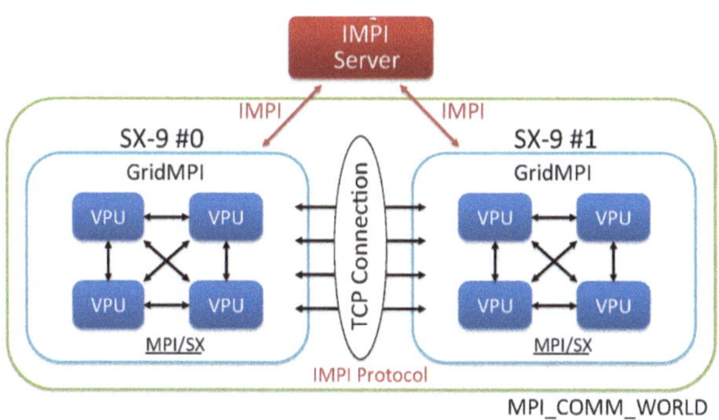

Fig. 7 MPI operation environment

4 Feasibility Study and Early Performance Evaluations

To confirm the potential of the vector computing cloud, this chapter presents early performance evaluation and operation tests of the prototype system. First, the performance of proposed job scheduling mechanism is evaluated by simulation and operation tests. Then, HPL benchmark is used to clarify the potential of the vector computing cloud.

4.1 Performance Evaluation of the Job Scheduling Mechanism

4.1.1 Simulation Analysis

To evaluate the performance of the proposed job scheduling mechanism, the vector computing cloud environment composed of two queuing-based sites is modeled and simulated. This analysis is carried out by using ten kinds of jobs, and the job-execution times of all jobs are generated in the gamma distribution with the averages (μ) and the standard deviation (σ) shown in Table 1. The ten kinds of jobs are generated with zipf's law [17]. By using the zipf's law, many small-jobs and few large-jobs are generated in the simulation. The zipf's law can be applied to many natural and social phenomena, and this evaluation assumes the zipf's law as a real HPC user's workload. In the initial phase of the simulation, thirty jobs are sequentially submitted to the vector computing cloud. Then, thirty jobs always exist in the vector computing cloud while simulation is executed. To evaluate this situation, whenever one job execution has been finished, a new job is generated and submitted. The proposed job scheduling mechanism and the *round-robin* scheduling mechanism used in the NAREGI Grid Middleware are evaluated. These simulations are carried out until 1,000,000 simulation seconds, and the period from the submission time to the start time of a job(*waiting-time*), and the number of jobs which is completed are measured. The evaluated results are obtained by taking the average of twenty simulations.

Table 2 summarizes the simulation results. From this result, the proposed job scheduling mechanism improves the number of executed jobs and reduces the average and maximum waiting-time. The standard deviation of waiting-time shows that the waiting-time of jobs scheduled by the proposal concentrates on the average, but the waiting-time of jobs scheduled by round-robin is distributed over the wide region.

Figure 8 shows the histograms of the waiting-time. The horizontal axis indicates the waiting-time of jobs, and the vertical axis is the number of jobs that are within the waiting-time. Figure 8 shows that some jobs scheduled by the round-robin scheduler have been executed as soon as it is submitted. Because of the simulation condition that thirty jobs are always allocated to two sites, a case of the

Table 1 Simulation parameters

Parameter name	Value
average execution time of each job (μ) [sec]	100, 200, 400, 800, 1600, 3200, 6400, 12800, 25600, 51200
standard deviation of each job (σ)	11

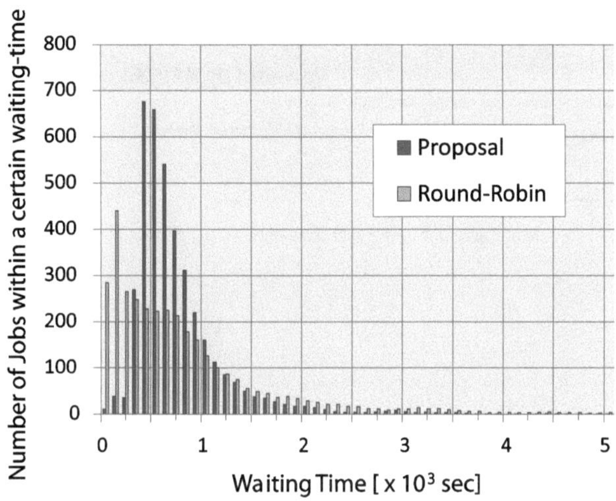

Fig. 8 Histogram of waiting-time (1,000 second intervals)

Table 2 Summary of simulation results

	Proposal	Round-robin
number of executed jobs	3,914	3,458
average of waiting-time [sec]	7,131	8,085
maximum of waiting-time [sec]	81,464	94,113
standard deviation of waiting-time	6,477	10,368

waiting-time being zero indicates the situation that all jobs are allocated to the one site, and no job is allocated to the another site. The round-robin scheduling makes the load-imbalance among two sites, and only a part of the entire computing power in the vector computing cloud is utilized. This load-imbalance by the round-robin scheduling causes the low number of executed jobs in Table 2.

On the other hand, the proposed job scheduling mechanism can provide the fair waiting-time for all jobs and eliminate the load-imbalance among two sites. Then, the proposed job scheduling mechanism improves the utilization efficiency of the computing resources in the vector computing cloud.

4.2 System Tests

To confirm the effectiveness of proposed job scheduling algorithms and the usability of the vector computing cloud, experiments are carried out. In the vector computing cloud, users are logged in to geometrically distributed vector computing resources from a server as shown in Fig. 9. Then operating commands are put on a job submission page of the portal site as shown in Fig. 10. In this page, users can also confirm the resource availability of the vector computing cloud by a signal icon.

Fig. 9 Portal site page: login

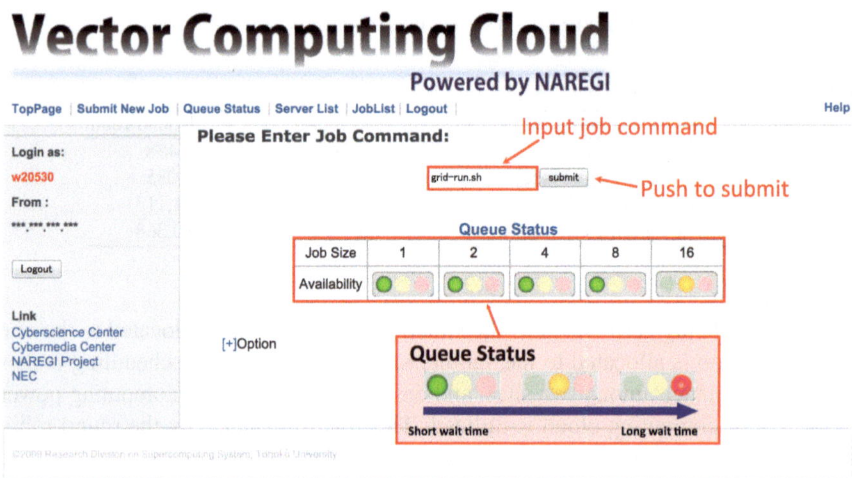

Fig. 10 Portal site page: submit

The resource availability is obtained by IS server, and a waiting time for users jobs are displayed with respect to each job queue. From the user's point of view, both systems of Tohoku University and Osaka University are completely virtualized as a single system, and user do not have to care where users job should submit or will be executed. In addition, users can also confirm more detailed system load information from the portal page as shown in Fig. 11.

In this operation test, jobs with different job-execution time are successively submitted to the prototype system in keeping with the "submission order" in Table 3. The sequence of jobs consists of three kinds of applications, and their execution times are set to 30 (small), 40 (middle) and 200 (large) seconds, respectively. All submitted jobs are satisfactorily allocated and executed. The job allocation results

Table 3 List of submitted jobs

Submission order	Job ID	Job size
1	CID_432	small
2	CID_433	middle
3	CID_434	large
4	CID_435	middle
5	CID_436	small
6	CID_437	small
7	CID_438	small
8	CID_439	middle

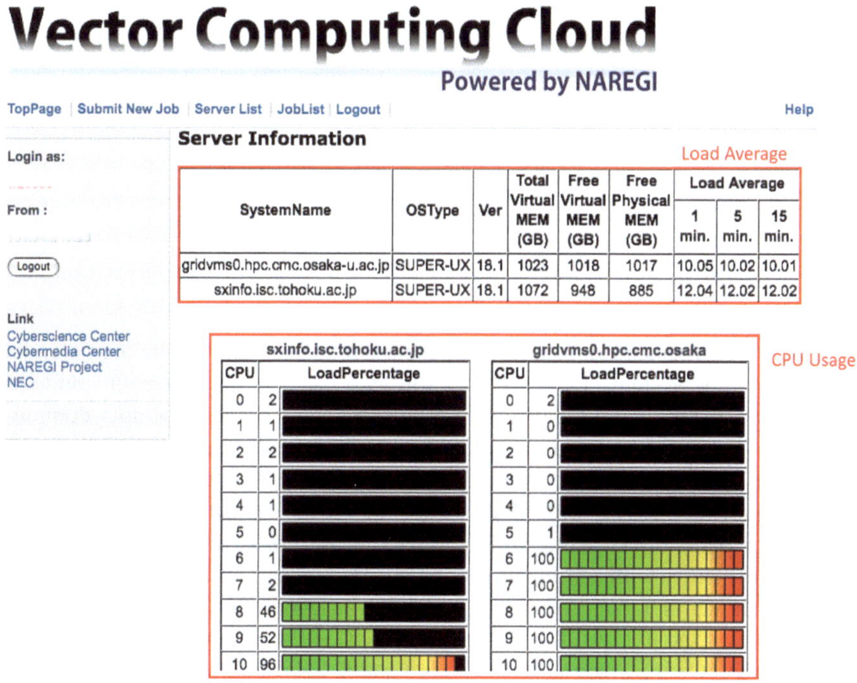

Fig. 11 Portal site page: system load

are confirmed by a portal site as shown in Fig. 12. In this screenshot, the first row indicates allocated jobs to Tohoku University's site and the second row indicates those of Osaka University's site. The horizontal axis is the time sequence. In this result, after job **CID_434** which has the longest execution time is allocated to Tohoku University, the job-start time of Tohoku University becomes much longer than that to Osaka University. As a result, the jobs **CID_435–CID_438** are sequentially allocated to Osaka University to make quick response to users and reduce the imbalance of job-start time between both sites. From the results, the job scheduler can well estimate the execution time of each job, and select the appropriate site to execute the job as early as possible in the vector computing cloud that consists of computing resources with different job operating policies.

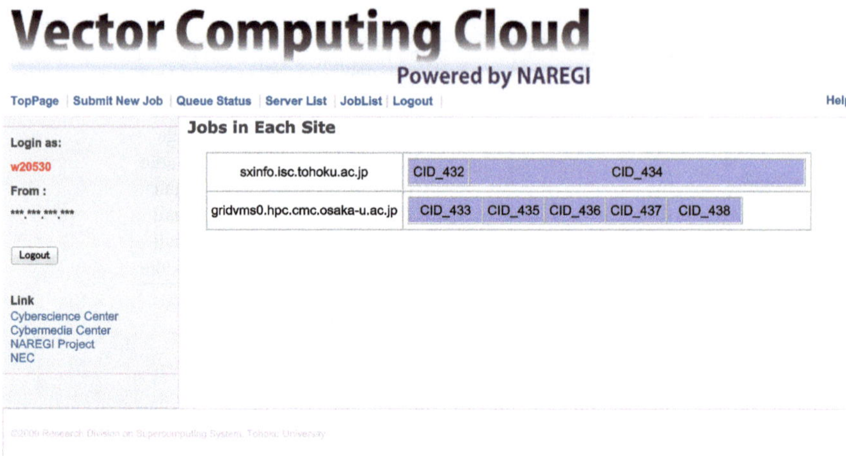

Fig. 12 Portal site page: result of job scheduling

4.3 Performance of HPL

The High Performance Linpack (HPL) benchmark is used to evaluate the potential of the prototype system [18]. HPL is a software package that solves a (random) dense linear system in double precision arithmetic on distributed-memory computers. It can thus be regarded as a portable as well as freely available implementation of the High Performance Linpack benchmark. For this evaluation, two nodes of SX-9 with 16 CPUs and the HPL benchmark with standard "increase 2-ring" topology on panel broadcasting with a 32 parallel MPI program are used. We have evaluated five cases with combinations of network connections and MPI implementations as shown in Table 4. In Cases 1 and 2, SX-9 nodes at Osaka university are connected by IXS at 128 GB/s (bi-directional). Case 1 uses MPI/SX and Case 2 uses GridMPI. On the other hand, in Cases 3 and 4, SX-9 nodes at Osaka university are connected through 1 GbE with GridMPI. Case 3 uses jumbo flame TCP/IP communication (9000 bytes/flame), and Case 4 employs normal flame TCP/IP communication (1500 bytes/flame). Case 5 is running HPL on the vector computing cloud with GridMPI as shown in Fig. 7. SX-9s at Tohoku University and Osaka University are connected through SINET3, and the round trip time (RTT) between two SX-9 systems at Touhoku University and Osaka University is 24 ms.

The experimental results are shown in Fig. 13. In this experiment, the problem size N is varied from 9,000 to 144,000. As shown in the results, as the problem size becomes large, the performance increases. The reason is that the communication overheads are hidden by computations as the problem size increases. Case 1 indicates the ideal case to operate the benchmark and realize the highest computation efficiency. On the other hand, due to the protocol transformation from GridMPI to MPI/SX, the performance of Case 2 is decreased compared to the Case 1. The performances of Case 4 are lower than that of Case 2. The results emphasize the

Table 4 Experimental setup

Name	Node #0 located at	Node #1 located at	Node interconnect	MPI implementation
case1	Osaka University	Osaka University	IXS	MPI/SX
case2	Osaka University	Osaka University	IXS	GridMPI
case3	Osaka University	Osaka University	GbE with Jumbo flame	GridMPI, IMPI
case4	Osaka University	Osaka University	GbE with Normal flame	GridMPI, IMPI
case5	Osaka University	Tohoku University	SINET 3(GbE RTT=24 ms)	GridMPI, IMPI

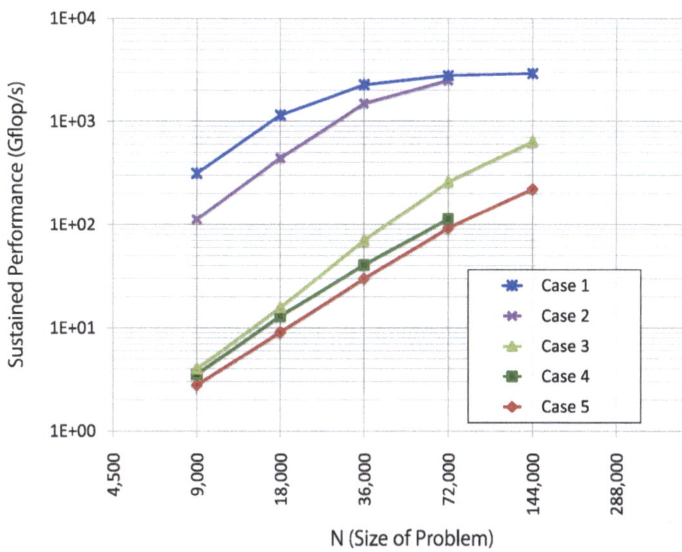

Fig. 13 HPL performance

effect of network performance, because the difference between Cases 2 and 4 is just network connection between each SX-9. Although the performance of Case5 is scaled as N increases, it shows the lowest performance in the all cases. Let focus on Cases 4 and 5, the difference between these cases is only network interface as Cases 2 and 4. From these facts, we can re-confirm the importance of the network performance of each node. To put it the other way around, there is room to improve the performance of the vector computing cloud as useful HPC cloud infrastructure by improving the network performance. For example, the results of Case 3 indicate the effectiveness of the jumbo flame TCP/IP communication. In addition, we can also improve the network performance by employing a 10 GbE or TCP/IP offloading Engines. From these results we can confirm that the vector computing cloud has enough potential to achieve higher performance than that of Cases 3 and 4 by introducing network performance enhancement technologies. These challenges and their performance evaluations are remained as our future work.

5 Conclusions

This paper has presented the world's first prototype of wide-area vector meta-computing infrastructure named the vector computing cloud. The prototype system, which is designed based on the NAREGI Grid middleware has successfully realized a single sign on environment by virtualizing geographically distributed vector supercomputers as an HPC service over the Internet. To enhance the job and resource management capabilities of the NAREGI Grid Middleware, GridVM for SX and job scheduler have been introduced and developed. The proposed job scheduling mechanism obtains the job-start time in a queuing-based system from the history of the job-execution times, and automatically allocates a job to an appropriate site, which can execute the job earlier. The experiment results indicate that the proposed job scheduling mechanism has enough potential for transparently operating jobs between the two SX-9 systems with coexistence of conventional jobs and cloud jobs. Furthermore, though the RTT between remote computers was 24 ms, the HPL benchmark results have shown an enough potential for the future vector meta-computing infrastructures.

In our future work, we will increase a number of nodes by involving other computer centers to the vector computing cloud. We will also try to improve the network performance by introducing 10 GbE and ToE and an intelligent network technology to bring up our system as useful HPC cloud infrastructure. In addition, improving the capabilities of the job scheduler and the efficiency of GridMPI is needed to realize effective usage of the vector computing cloud.

Acknowledgements This work was partially supported National Institute of Informatics (NII). The author would like to extend our thanks to the members of the Center of GRID Research and Development (NII), and Kenji Oizumi, Eiichi Ito of Cyberscience center Tohoku University, and Masa-aki Yamagata, Norio Kamiyama, Hironobu Konno of NEC Corp. for their help to implement the prototype system.

References

1. The Globus Alliance. http://www.globus.org/.
2. A. Weiss. Computing in the clouds. *netWorker*, 11(4):16–25, 2007.
3. B. Hayes. Cloud computing. *Communications of the ACM*, 51(7):9–11, 2008.
4. B. Sotomayor, K. Keahey, and I. Foster. Combining batch execution and leasing using virtual machines. In *HPDC '08: Proceedings of the 17th international symposium on High performance distributed computing*, pages 87–96, New York, NY, USA, 2008. ACM.
5. B. Sotomayor, R. Montero, I. Llorente, and I. Foster. Virtual infrastructure management in private and hybrid clouds. *IEEE Internet Computing*, 13:14–22, 2009.
6. Takashi Soga, Akihiro Musa, Chichi Shimomura, Ryusuke Egawa, Ken'ichi Itakura, Hiroyuki Takizawa, Koki Okabe, and Hiroaki Kobayashi. Performance evaluation of nec sx-9 using real science and engineering applications. In *SC '09: Proceedings of the Conference on High Performance Computing Networking, Storage and Analysis*, pages 1–12, New York, NY, USA, 2009. ACM.

7. Leonid Oliker, Jonathan Carter, Michael Wehner, Andrew Canning, Stephane Ethier, Art Mirin, David Parks, Patrick Worley, Shigemune Kitawaki, and Yoshinori Tsuda. Leading computational methods on scalar and vector hec platforms. In *SC '05: Proceedings of the 2005 ACM/IEEE conference on Supercomputing*, page 62, Washington, DC, USA, 2005. IEEE Computer Society.
8. Amazon Inc. Amazon web services ec2 pages. *sdkoakosdaksopda.*
9. Google Inc. Google apps pages. http://www.google.com/apps/intl/en-GB/business/index.html.
10. Grid Technology and Research Center. Gridmpi pages. http://www.gridmpi.org/gridmpi.jsp.
11. Center for Grid Research and Development. National research grid initiative:naregi pages. http://www.naregi.org/index_e.html.
12. National Institute of Informatics(NII). Next-generatiion science information network 3:sinet3 pages. http://www.sinet.ad.jp/?set_language=en.
13. Toshiyuki Kitagawa, Shoichi Hasegawa, and Akihiro Yamashita. Job scheduling function nqs ii for sx-6. *NEC Technical Journal*, 55(9):50–53, 2002/9.
14. Manabu Higashida. *High Performance Computing on Vector Systems 2009*, chapter The Grid Middleware on SX and Its Operation for Nation-Wide Service, pages 109–119. Springer Berlin Heidelberg, 2009.
15. Matthew L. Massie, Brent N. Chun, and David E. Culler. The ganglia distributed monitoring system: design, implementation, and experience. *Parallel Computing*, 30(7):817–840, 2004.
16. William L. George, John G. Hagedorn, and Judith E. Devaney. Impi: Making mpi interoperable. *Journal of Research of the National Institute of Standards and Technology*, 105:343–428, 2000.
17. George K. Zipf. Human behavior and the principle of least effort. Addison-Wesley, 1949.
18. A. Petitet, R. C. Whaley, J. Dongarra, and A. Cleary. Hpl—a portable implementation of the high-performance linpack benchmark for distributed-memorycomputers. http://www.netlib.org/benchmark/hpl/.

Leonid Oliker, Jonathan Carter, Michael Wehner, Andrew Canning, Stephane Ethier, Art Mirin, David Parks, Patrick Worley, Shigemune Kitawaki, Yoshinori Tsuda. Leading computational methods on scalar and vector HEC platforms. [illegible]

Full-Scale 3D Vibration Simulator of an Entire Nuclear Power Plant on Simple Orchestration Application Framework

Guehee Kim, Kohei Nakajima, Takayuki Tatekawa, Naoya Teshima, Yoshio Suzuki, Hiroshi Takemiya

Abstract So far, we have developed grid-enabled application for "Full-Scale 3D Vibration Simulator for an Entire Nuclear Power Plant" which is simulation platform to analyze seismic response of a whole digitalized nuclear power plant. In the 3D Vibration Simulator, components of a nuclear power plant are treated in hierarchical manner in which large components are grouped at primary level and small components such as pipes are grouped at secondary level and boundary condition data from the large components simulation are used as input data of small components simulation. In this work, to make the whole simulation more efficient than the previous sequential scenario in which after large components simulation is completed, small components simulation starts, we introduce pipelined data-transfer scenario in which boundary condition data are transferred each time step while all components simulation are run in parallel. In realization of the 3D Vibration Simulator in the introduced scenario, we confronted two challenges: first, clearance of job's idle time to be wasted for only waiting data which takes from a few ten minutes to a few hours per each time step and second, immediate resubmission of abnormal ended jobs for a long time simulation under the introduced scenario. To address these challenges, we proposed two solutions: as first solution, we set policy by which jobs of small components are submitted after all necessary data per each time step arrive and executed only that time step, which process is repeated whenever next time step input data arrive and as second solution, we make an abnormal ended job automatically resubmitted. Since there were no pre-existing grid technologies to provide sufficient functionalities to enable these solutions to be possible from the previous grid-enabled application, we developed Simple Orchestration Application Framework (SOAF) and upgraded the previous grid-enabled application by implement-

Guehee Kim, Nakajima Kohei, Takayuki Tatekawa, Naoya Teshima, Yoshio Suzuki, Hiroshi Takemiya
Center for Computational Science and e-Systems, Japan Atomic Energy Agency, 6-9-3 Higashi-Ueno, Taito-ku, Tokyo 110-0015, Japan
e-mail: kim.guehee@jaea.go.jp, nakajima.kohei@jaea.go.jp, tatekawa.takayuki@jaea.go.jp, teshima.naoya@jaea.go.jp, suzuki.yoshio@jaea.go.jp, takemiya.hiroshi@jaea.go.jp

M. Resch et al. (eds.), *High Performance Computing on Vector Systems 2010*, 93
DOI 10.1007/978-3-642-11851-7_7, © Springer-Verlag Berlin Heidelberg 2010

ing the SOAF. Using the upgraded grid-enable application, we performed seismic analysis of High Temperature Engineering Test Reactor at O-arai R&D center of Japan Atomic Energy Agency and confirmed that the simulation were performed in pipelined data-transfer scenario effectively using computing resources without idle time for about a week simulation period resubmitting abnormally ended jobs. In this paper, the details of all of this work will be described.

1 Introduction

In Japan of earthquake-prone country, earthquake resistance design and dynamic analysis of nuclear power plants (NPPs) are very important issues. Japan has over 50 NPPs and over 10 reactors among them are aging as more than 30 years old. Further, the Great Hanshin Earthquake which devastated the city of Kobe in 1995 and Chūetsu offshore earthquake in 2007 have heightened people's concern about the safety of NPPs. Thus, keeping of safe operation of NPPs becomes all the more important task, which contributes to stable atomic energy supply making people feel reliable.

Center for Computational Science and e-Systems of Japan Atomic Energy Agency (CCSE/JAEA) has performed R&Ds to satisfy the above needs and developed "Full-Scale 3D Vibration Simulator for an Entire Nuclear Power Plant". Full-Scale 3D Vibration Simulator is full-fledged simulation platform for seismic response analysis for a whole NPP as shown in Fig. 1 [1]. Since the 3D Vibration Simulator treats a whole digitalized NPP as an assembly structure composed of its digitalized parts, it is to elucidate physical behaviors boundary between NPP parts in detail, which is impossible to research with present mass-spring model. Clarifying seismic performance of a whole NPP structure, we aim to help soundness evaluation of the existing NPPs and support desirable earthquake resistance design of new NPPs and upgrading of the design concept.

Generally, full-fledged simulation of an NPP using 3D Vibration Simulator needs high computing power and memory capacity that a present single supercomputer can not sufficiently afford. In future, it is expected that tera- or peta-scale computing resources should be needed in all respects of memory, computing performance, and storage. For this reason, R&Ds of parallel and distributed computing technologies which enable to couple various supercomputers geographically dispersed is very important issue for the realization of the full-fledged simulation. In this R&Ds, use of grid computing technology [2, 3] is inevitable and in this context, we have developed grid-enabled application of the 3D Vibration Simulator using AEGIS (Atomic Energy Grid InfraStructure) Client APIs (Application Program Interfaces) [4, 5].

AEGIS developed and established by CCSE/JAEA provides supercomputing resources of various universities, computational science facilities, institutes, and so on in Japan [4–9]. In AEGIS, main targeted simulations are large-scale simulations in atomic energy field, for example, burning plasma [15] and prediction of quake-proof

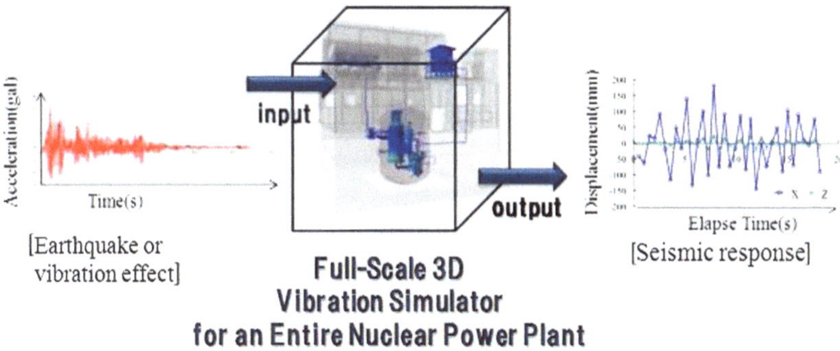

Fig. 1 "Full-Scale 3D Vibration Simulator for an Entire Nuclear Power Plant" is full-fledged simulation platform for seismic response analysis for a whole digitalized nuclear power plant

capability of NPPs as well as 3D Vibration Simulator. So far, we have supported simple and easy development of grid-enabled applications for these simulations by providing AEGIS Client APIs. Installing the application on user's terminal, users can perform simulations feeling like as they use desktops with supercomputing power without consciousness of complexity of grid service layers.

In 3D Vibration Simulator, an NPP model is deployed on computing resources by component unit which is composed of a number of parts. In previous grid-enabled application (hereafter, GDS (Grid-enabled Desktop Supercomputing) application) [4, 5], all of large components such as reactor vessel are first simulated, after that, using output data from the large components simulation, small components, for example, interconnecting pipes were simulated. In this work, we introduce pipelined data-transfer scenario to make the whole simulation more efficient, by which all components are simultaneously simulated transferring data per each time step from large components to small components. However, we confronted two problems: first problem is waste of computing resources due to existence of idle time to wait arrival of necessary input data from the large components which took even a few hours sometime per each step and second problem is needs of immediate resubmission of abnormal ended jobs for realization long time simulation under pipelined data-transfer scenario. Since pre-existing grid technologies did not provide sufficient functionalities to simply adapt these solutions in the grid-enabled application, we developed Simple Application Orchestration Framework (SOAF) [15] and implementing the SOAF in the previous grid-enabled application of 3D Vibration Simulator. Using the upgraded grid-enable application we performed seismic response analysis of High Temperature Engineering Test Reactor (HTTR) at O-arai R&D center of JAEA and confirmed that the proposed solutions in this work were very effective so that long time simulation of about 160 hours by well functioned automatic resubmission of abnormal ended jobs was realized without wasteful use of computing resources in pipelined data-transfer scenario. In Sect. 2, we explain 3D Vibration Simulator and discuss the above two problems and their solutions in detail. In Sect. 3, we describe SOAF functionality and its implementation. In Sect. 4, details of seismic response simulation of HTTR will be explained. In Sect. 5, we summarize our work.

2 Full-Scale 3D Vibration Simulator for an Entire Nuclear Power Plant

2.1 GDS Application of Full-Scale 3D Vibration Simulator

Distinguished point of 3D Vibration Simulator is that a whole digitalized NPP is divided by component units, where each component is an assembly structure with a number of NPP parts prepared in mesh model of finite elements method as shown in Fig. 2. Components are individually analyzed by 3D Vibration Simulator solver, FIESTA (FInite Elements STructural Analysis) [1] and mutual vibration effects between parts are reflected by data exchange at parts boundaries per each simulation time step. 3D structural analysis by FIESTA elucidates detailed mechanism of physical phenomena occurred at parts boundaries, which is very important to guarantee soundness of NPPs but impossible by de-facto mass-spring modeling methodology.

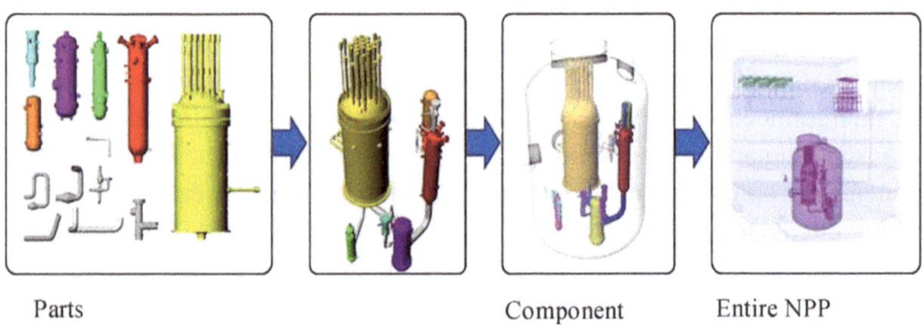

Fig. 2 In 3D Vibration Simulator, an NPP is considered of assembly structure composed of parts in finite element method model

An entire NPP is composed of parts from 100 thousands to 10 millions so that generally, it is impossible to perform full-scale simulation on a single supercomputer. In future, as analysis precision will become more and more accurate, prepared model data will be terabyte-scale size and we will need tera- or peta-scale resources for computing performance, memory, storage, and so on. In this respect, use of grid resources is desirable solution, and by deploying components on various grid computing resources we can realize 3D Vibration Simulator methodology of "from parts to the whole".

CCSE/JAEA has more advanced grid technologies acquired from R&D of ITBL (IT-Based Laboratory) Project [10–13] and established grid infrastructure, AEGIS for atomic energy research field. So far, we have provided AEGIS Client APIs which support AEGIS users to develop their own GDS application in a simple and easy way. We aim that GDS applications provide simulation environment which makes users feel like as they have their own desktop supercomputers. As shown in Fig. 3, using the Client APIs, we developed GDS application of 3D VPVS and have performed seismic analysis of an NPP which was reported in the 9th Teraflops Workshop in 2008.

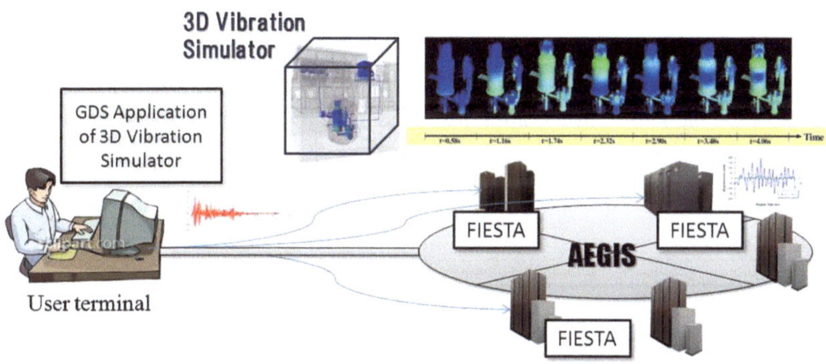

Fig. 3 CCSE/JAEA has developed GDS applications for 3D Vibration Simulator which connected with computing resources on AEGIS infrastructure

2.2 Needs of Pipelined Data-Transfer Scenario

Large-mass components of NPPs, for example, reactor pressure vessel and cooling systems are interconnected by small-mass components such as pipeline systems. In previous theoretical study [14], it was shown that physical effects from small components to these large components could be neglected compared to that the vice versa. Based on this result, we introduced hierarchical manner as shown in Fig. 4 by which we grouped large components at primary level and small components at secondary level and used boundary condition data obtained from simulation at primary level as input data to simulate secondary level.

In this work, we introduce pipelined data-transfer scenario in which boundary condition data are transmitted per each time step from the primary level to the secondary level. Pipelined data-transfer makes the whole simulation more efficient than the previous sequential data-transfer scenario as shown in Fig. 5. In sequential data-transfer, boundary condition data are accumulated during simulations at primary

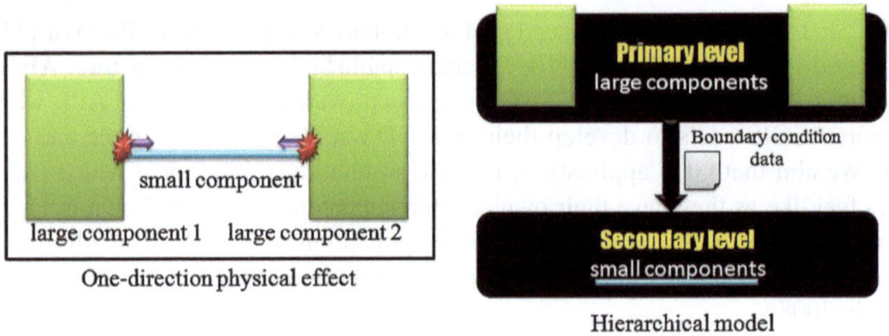

Fig. 4 In 3D Vibration Simulator, considering one-direction physical effect, all components are grouped into primary level of large components and secondary level of small components

level and after the simulation completes, used as input data with the start of simulations at secondary level.

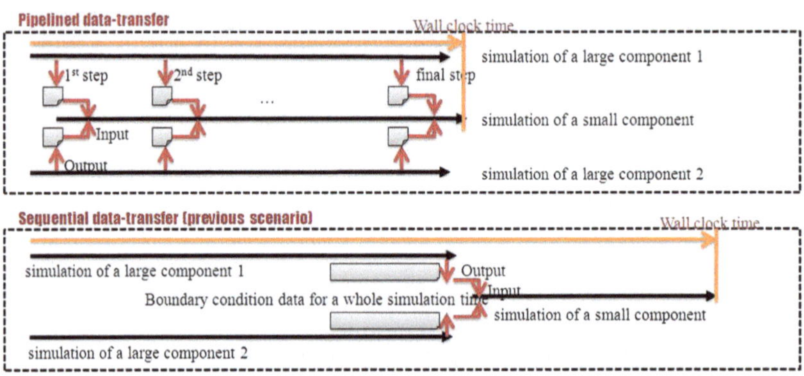

Fig. 5 Pipelined data-transfer makes a whole simulation more efficient than sequential data-transfer

In adopting the pipelined data-transfer scenario to the 3D Vibration Simulator, we confronted two challenges: first challenge is effective use of computing resources and second challenge is realization of long time simulation adopting pipelined data-transfer scenario through a whole simulation. In 3D VPVS, batch queue jobs at secondary level in idle state wait arrivals of all boundary condition data from primary level and the waiting time takes from a few 10 minutes to a few hours per each time step. Problem is that the existence of the idle time is an obstacle which prohibits a whole effective use of computing resources which are shared by many researchers. On the other hand, to get physically useful analysis results, simulation needs a few weeks which exceed very far time limit assigned to batch queue jobs. Problem is that when even one of jobs at primary level is killed by a batch system, for example, for the reason of time limit exceed, the whole simulation can't step further. When even one of jobs at secondary level is abnormally ended, depending on the resubmit timing, it is difficult to realize pipelined data-transfer.

To address these challenges, we thought solutions as follows. As solution for the first challenge, we introduce "one time step execution" policy by which each job at secondary level is submitted after all of its one time step input data arrive and executed only that time step, which is repeated whenever new time step input data arrive. In this policy, since jobs don't have to wait data in idle state, first problem is cleared. As solution for the second challenge, monitoring each job status and immediately resubmit a job which abnormally is ended including the case that the jobs is killed by time limit exceed. By this immediate resubmit, simulation can be executed under pipelined data-transfer for long time simulation.

3 Development of Simple Orchestration Application Framework

The previous GDS application lacked functionalities to realize the pipelined data-transfer scenario adapting two solutions proposed in Sect. 2.2 because the application was designed for the sequential data-transfer scenario. The application permitted only sequential control of data-transfer and job execution so that data could be transferred only after a job was completely terminated and there was no way to transfer the data during job's running. Besides, although the application regularly checked each job's status obtained from the AEGIS Client APIs, that was only for user's monitoring and when a job's abnormal end was informed, users needed resubmit the abnormal ended job themselves. To immediately resubmit a job, users must always keep their eyes on the monitoring messages.

To realize pipelined data-transfer scenario with these the solutions of "one time step execution" policy and job's immediate resubmit, we developed SOAF. In designing SOAF, we intended it to be general-purpose to also enable various other type scenarios necessary for other scientific simulations which couple multiple codes installed on distributed computing resources. For example, multi-physics or multi-scale simulations are composed of various scientific codes and each code is used with dedicated purpose for a specific physical phenomena or space-time scale. Coupling these codes involves data transfer between jobs in pipelined scenario, conditional branch scenario or even the mixed type of all these scenarios. It means that many scientific simulations with sophisticated coupling scenarios can be realized by controlling each scenario's data flow. Focusing on this point, we designed and developed SOAF and also considered that its implementation should be simple and use of its functionalities should be easy.

3.1 Functionalities

SOAF enables various computational simulations composed of multiple codes to be executed in distributed or parallel in simple and easy way. SOAF focuses only data flow between jobs and, according to the flow, controls job-submit timing and data-transfer, which is done by collaboration between "controller" and "sentinels" of SOAF composites. Controller is a C language program installed on user terminal and sentinels are simple Perl scripts of send_sentinel, recv_sentinel, and void_sentinel on computing resources. Three sentinels consist of one set and the set must be located in the same directory with simulation codes. Users edit job attributes and data flow (hereafter, file flow) in configuration text file. Job attributes for a job are job name, computer name, code path, job directory, and so on. Unit of file flow is prepared per each file and has information of a filename, job name outputting it, list of job name receiving it, and so on.

With the process start of the controller, it submits jobs and simultaneously, sentinels enter to their monitoring processes. Each send_sentinel keeps watching whether its target file's flag file exists or not. Here, a flag file is an empty file generated by a simulation code after the target file is closed. When the flag file exists, the send_sentinel outputs a signal file and terminates its process (but, new process starts immediately for next step target file). After confirming the process termination of the send_sentinel, the controller transfers first, the target file and next, the signal file. After confirming arrival of the signal file, recv_sentinel generates a flag file with the same file name as the target file's flag file. In SOAF, as one of job attributes, users can define job type depending on a submit condition. In the case that a job needs output data from another job as a submit condition, the job is defined as type 0. In the case that a job needs to be submitted from the controller's start, the job is defined as type 1. Therefore, the controller submits type 0 after target file transferred is completed. Once the type 0 job starts to run, it confirms its target file's flag file and then, reads data of the target file. Here, we must notice that simulation source codes need to be added with a few code lines to enable output and confirmation of flag files. This is, however, trivial and very simple work compared to that needed when implementing other existing grid technologies [15]. Also, by the use of flag files and signal files, data output and data transfer are completely guaranteed. On the other hand, to immediately resubmit each abnormal ended job, the controller keep watch each job status and, when a job abnormally ends, automatically resubmits the job.

3.2 Implementation

We implemented SOAF in the previous GDS application of 3D Vibration Simulator. AEGIS Client APIs are located between controller and AEGIS middleware on computing resources as shown in Fig. 6. Job attributes in configuration file are passed to the Script Generator API, a middle level AEGIS Client API generating grid-enabled scripts of each job. Whenever the controller demands job submits, the Script Generator API hands over the job's scripts to AEGIS middleware via low level AEGIS Client APIs. When a job is killed by a batch system for time limit exceed, AEGIS Client APIs report returned value of job status, SCR_STATUS_ABNORMAL_END and accepting this report, controller resubmits the job.

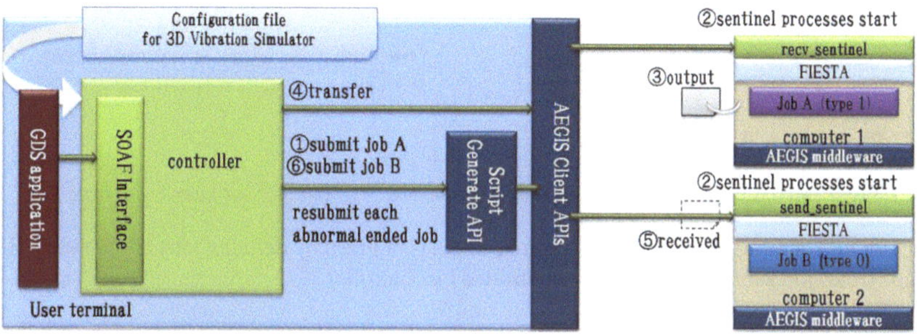

Fig. 6 View of system adapting SOAF in GDS application of 3D Vibration Simulator: file transfer between two jobs and resubmit of jobs are done by collaboration between controller and sentinels of SOAF composites via AEGIS Client APIs

As shown in Fig. 7, to adopt "one time step execution" policy in Sect. 2.2, we defined jobs at primary level as type 1 and jobs at secondary level as type 0. Jobs at primary level are submitted from the start of simulation and generate files having boundary condition data necessary as input data of jobs at secondary level. Whenever these files are generated, they are transferred to the job directories at secondary level and after that, jobs at secondary level are submitted. To enable jobs to advance further simulation time step through resubmission, we recoded simulation code so that restart files of jobs at primary level are generated at fixed interval time steps and those of jobs at secondary level are generated each time step.

4 Full-Scale Simulation of High Temperature Test Engineering Reactor

Using the upgraded GDS application by implementation of SOAF, we performed full-scale seismic response analysis of High Temperature Engineering Test Reactor (HTTR) at O-arai R&D center of JAEA in pipelined data-transfer scenario. HTTR, bird's-eye view of which is shown in Fig. 8(left), is research facility for development of high temperature gas-cooled reactor technology and nuclear heat utilization technology. Digitalized model of HTTR as shown in Fig. 8(right) is composed of six components: RPV (Reactor Pressure Vessel), PWAC (Pressurized Water Air Cooler), and AWAC (Auxiliary Water Air Cooler) at primary level and three interconnecting pipe systems at secondary level. Total degree of freedom is about 150

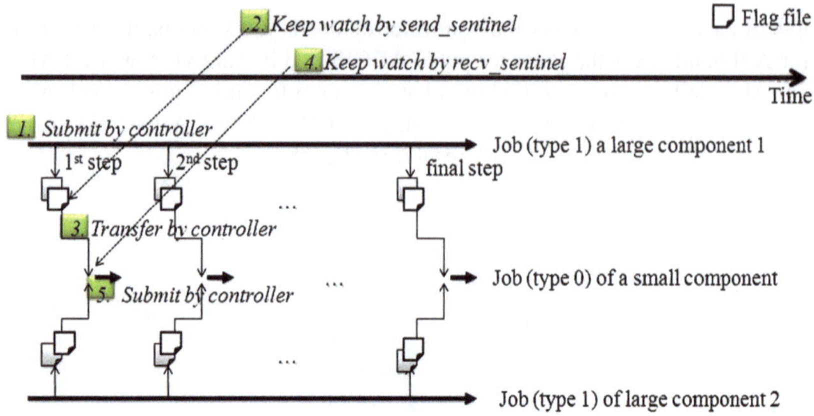

Fig. 7 "One time step execution" policy in Sect. 2.2 is realized by defining jobs at secondary level as type 0 and job submission timing is controlled by controller

million and total data size is about 26 GB. Table 1 shows number of mesh nodes and model data size of each component. We used 20 seconds data of El Centro earthquake in 1940 to perform full-scale seismic response analysis of HTTR which demanded 1 terabytes memory.

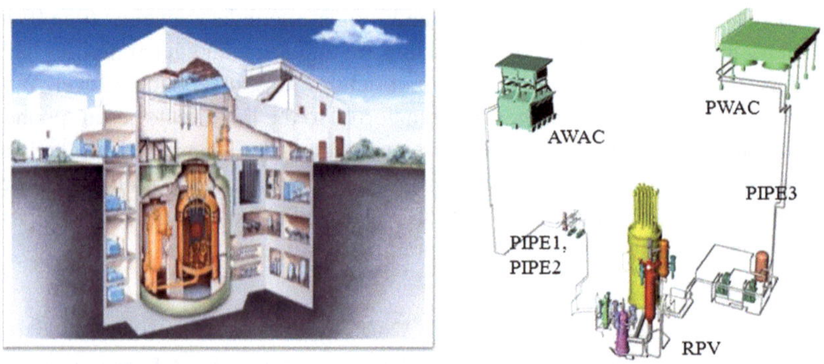

Fig. 8 Bird's-eye view of HTTR building (left side) and its digitalized model composed of six components (right side)

We used a massive parallel supercomputer, Altix3700Bx2 and a PC cluster machine, Altix350 on AEGIS Tokai site of JAEA as shown in Fig. 9. Total number of CPUs was 1024 using six batch queue job classes as shown in Fig. 10. Considering that it is not desirable to use of plural number of data files as input files for simulation efficiency, we prepared a TSS (Time Sharing System) program between primary level and secondary level to unify the plural files from the primary level into one file.

Table 1 Details of six digitalized components of HTTR

Level	Components	Number of mesh nodes	Mesh data size (GB)
Primary	RPV	26,047,774	16.2
	PWAC	13,810,729	5.9
	AWAC	3,364,149	1.4
Secondary	PIPE1	622,090	0.2
	PIPE2	1,830,427	0.8
	PIPE3	3,396,149	1.7
	Total	49,071,318	26.2

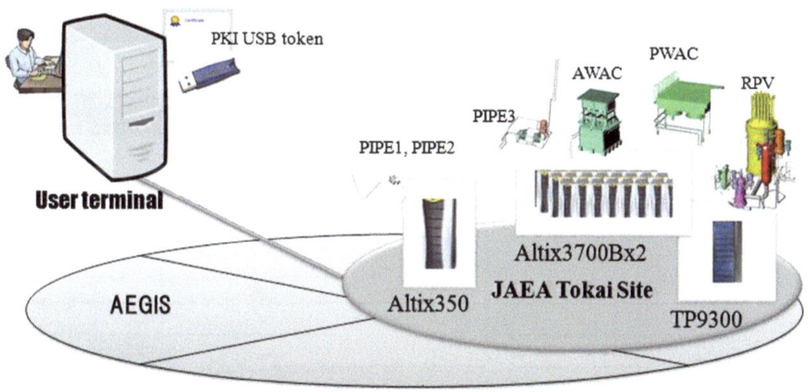

Fig. 9 Deployment view of six components of HTTR on computing resources in JAEA Tokai site of AEGIS

We confirmed effectiveness of the proposed solutions in Sect. 2.2 through this full-scale simulation of HTTR. During the whole simulation time, all of data files were transferred to their target directories in pipelined scenario. Since there were jobs of type 0 by the TSS program, pipelined data-transfer was done through two stages: from primary level to TSS level and from the TSS level to secondary level. By "one time step execution" policy, we could simulate without wasteful use of computing resources because there were no idle states. Immediate resubmissions of abnormally ended jobs due to time limit exceed were done one time for RPV, three times for PWAC, and 9 times for AWAC, where limit time of RPV job class was 12 hours and others were 3 hours for total simulation period of about 160 hours. In this simulation, there was no time limit exceed in the case of jobs at secondary level.

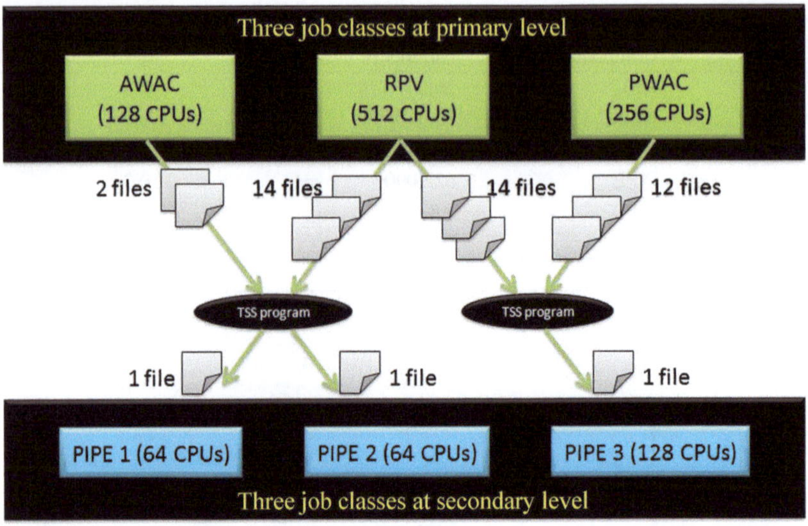

Fig. 10 File transfer view per each time step from three job classes at primary level to three job classes at secondary level via TSS programs which unifying plural files to one file

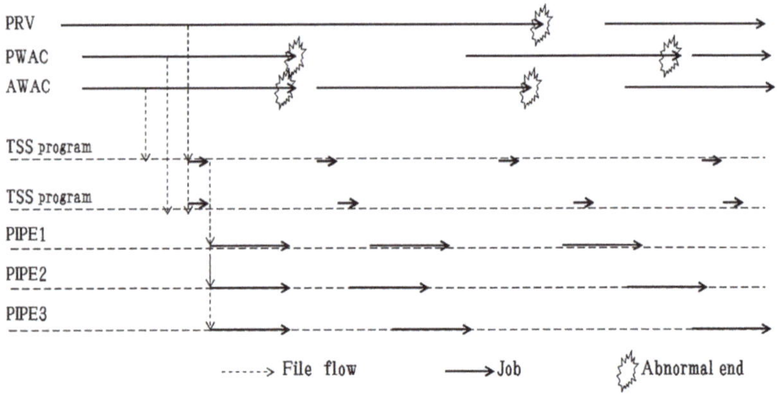

Fig. 11 Schematic view of jobs flow chart according to pipelined data-transfer scenario between six job classes and two TSS programs

5 Summaries

In Japan, earthquake resistance design and dynamic analysis of NPPs are very important issues. To address these issues, CCSE/JAEA has performed R&Ds of "Full-Scale 3D Vibration Simulator for an Entire Nuclear Power Plant" which is seismic response analysis system for a whole digitalized NPP. Generally, full-fledged simulation of an NPP needs massive computing power, very huge storage, memory, and so on which a single supercomputer could not sustain sufficiently. To tackle this situation, we have developed grid-enable application of 3D Vibration Simulator using AEGIS Client APIs of grid infrastructure AEGIS developed by JAEA and an NPP model is deployed on computing resources of AEGIS by NPP component unit. In NPP model of 3D Vibration Simulator, large components such as a reactor pressure vessel are interconnected by small components such as pipes and 3D Vibration Simulator treats physical effects between them as one-direction effects from large components to small components. In this work, to reflect the one-direction physical effect, we introduced pipelined data-transfer scenario in which boundary condition data were transferred from the large components to small components per each time step, which made simulation more efficient than the previous sequential scenario in which after completion of all simulation of large components, simulation of small components began. In adopting the pipelined data-transfer scenario, we confronted two challenges: effective use of computing resources and realization of long time simulation. To address these challenges, we proposed solutions: 1) "one time step execution" policy in which jobs for small components started after all necessary input data arrived and executed only one step and 2) immediate resubmission of abnormal ended job. Since the previous grid-enabled application lacked functionalities to adapt these solutions we developed Simple Orchestration Application Framework (SOAF). SOAF had functionality of file transfer and job submission depending on the only files flow between jobs and functionality of automatic resubmission of abnormally ended job. We implemented SOAF in the previous grid-enabled application, from which we performed full-scale seismic response analysis for High Temperature Engineering Test Reactor at O-arai R&D center of JAEA using six job classes of two supercomputers. Through this simulation which took almost a week, we confirmed the effectiveness of the pipelined data-transfer adapting the proposed solutions was executed without waste of computing resources and abnormal ended job were automatically resubmitted.

Acknowledgements We acknowledge O-arai R&D center of JAEA for useful discussion and data provision.

References

1. Yamada, T. and Araya, F. (2009) Construction of vibration table in an extended world for safety assessment of nuclear power plants. High Performance Computing on Vector, Springer, Heidelberg: 223–232
2. Foster, I., Kesselman, C. et al. (1998) The Grid: blueprint for a new computing infrastructure. Morgan Kaufmann, California
3. Foster, I., Kesselman, C. et al. (2004) The Grid 2: blueprint for a new computing infrastructure. Morgan Kaufmann, California
4. Kim, G., Suzuki, Y. et al. (2009) A Script Generator API for the full-scale three-dimensional vibration simulation of an entire nuclear power plant within AEGIS. Proceedings of The 1st International Conference on Parallel, Distributed and Grid Computing for Engineering (PARENG2009): DOI:10.4203/ccp.90.22

5. Kim, G., Suzuki, Y. et al. (2009) Development of APIs for desktop supercomputing. High Performance Computing on Vector Systems, Springer, Heidelberg: 97–107
6. Suzuki, Y., Nakajima, K. et al. (2008) Research and development of fusion grid infrastructure based on atomic energy grid infrastructure (AEGIS). Fusion Engineering and Design 83: 511–515
7. Suzuki, Y., Nakajima, N. et al. (2007) Development of three-dimensional virtual plant vibration simulator on grid computing environment ITBL-IS/AEGIS. Proceedings of the Sixteenth International Conference on Nuclear Engineering CDROM:48478
8. Kim, G., Suzuki, Y. et al. (2009) Network computing infrastructure to share tools and data in GNEP. Proceedings of the Seventeenth International Conference on Nuclear Engineering CDROM:75304
9. Kim, G., Suzuki, Y. et al. (2010) Network computing infrastructure to share tools and data in Global Nuclear Energy Partnership. Journal of Power and Energy System 4: 180–190
10. Higuchi, K., Imamura, T. et al. (2003) Grid computing supporting system on ITBL project. Proceedings of the Fifth International Symposium on High Performance Computing (LNCS2858): 245–257
11. Imamura, T., Yamagishi, N. et al. (2003) A visual resource integration environment for distributed applications on the ITBL system. Proceedings of the Fifth International Symposium on High Performance Computing (LNCS2858): 258–268
12. Fukui, Y., Stubbings, A. et al. (2003) Constructing a virtual laboratory on the internet: the ITBL project. Proceedings of the Fifth International Symposium on High Performance Computing (LNCS2858): 288–297
13. Tani, M., Nakajima, N. et al. (2007) A methodology of structural analysis for nuclear power plant size of assembly. Proceedings of Mathematics & Computation and Supercomputing in Nuclear Applications: the Joint International Topical Meeting CDROM
14. Nishida, A., Liu, P. et al. (2000) Fundamental studies of wave-propagation properties of single layer lattice structures. JSCE Journal of Structural Engineering 46B:175–179 (in Japanese)
15. Tatekawa, T., Nakajima, K. et al. (2010) Simple Orchestration Application Framework to control "Burning Plasma Integrated Code", Proceedings of The Third International Joint Conference on Computational Sciences and Optimization (CSO2010) 2:322–326

Development of Simple Orchestration Application Framework and Its Application to Burning Plasma Simulation

Takayuki Tatekawa, Kohei Nakajima, Guehee Kim, Naoya Teshima, Yoshio Suzuki, Hiroshi Takemiya

Abstract We have developed the Simple Orchestration Application Framework (SOAF) to cooperatively control simulation codes on remote computers from a client PC. SOAF enables researchers to cooperatively execute various codes on grid infrastructure by only describing a configuration file including the information of execution codes and file flows among them. SOAF does not need substantial modification of the simulation codes. We have applied SOAF to the "Burning Plasma Integrated Code" which consists of various plasma simulation codes to solve the current diffusion, stability of plasma, current drive, and so on. In order to predict and interpret the behavior of fusion burning plasma, it is necessary to integrate simulation codes for complex plasma phenomena with wide temporal and spatial ranges. Since those codes exist on distributed heterogeneous computers installed in different sites such as universities and institutes, a grid computing technology is needed to cooperatively execute those codes. However, traditional grid technologies are difficult for non-computer scientists to use. By using SOAF, we successfully execute four plasma simulation codes included in the "Burning Plasma Integrated Code" according to the scenario described in the configuration file.

1 Introduction

In order to predict and interpret the behavior of fusion burning plasma, it is necessary to simulate complex plasma phenomena with wide temporal and spatial ranges. So far, various plasma simulation codes such as transport codes, MHD codes, par-

Takayuki Tatekawa, Kohei Nakajima, Guehee Kim, Naoya Teshima, Yoshio Suzuki, Hiroshi Takemiya
Center for Computational Science and e-Systems (CCSE), Japan Atomic Energy Agency (JAEA), 6-9-3 Higashi-Ueno, Taito-ku, Tokyo, 110-0015 Japan
e-mail: tatekawa.takayuki@jaea.go.jp, nakajima.kohei@jaea.go.jp, kim.guehee@jaea.go.jp, teshima.naoya@jaea.go.jp, suzuki.yoshio@jaea.go.jp, takemiya.hiroshi@jaea.go.jp

M. Resch et al. (eds.), *High Performance Computing on Vector Systems 2010*, 107
DOI 10.1007/978-3-642-11851-7_8, © Springer-Verlag Berlin Heidelberg 2010

ticle codes, CD/Heating codes, and so on, have been developed in universities and institutes to simulate such complex phenomena. And, their integration has also been tried to cooperatively execute those codes for integrated prediction and interpretation of the plasma behavior. There has been a problem how to integrate the simulation code on distributed computers cooperatively.

A grid computing technology has been used to realize their cooperative execution. For example, a workflow tool such as Kepler [1], TME [2], and so on enables re-searchers to build and execute scientific scenario from a client PC. Researchers can promote the execution of simulation codes and can visualize analysis processes using a graphical user interface (GUI). Here, researchers can construct the executable model of cooperation by simple procedures such as drag and drop. A remote procedure call (RPC) enables researchers to cooperatively manage various simulation codes by developing an application with its functions. GridRPC [3] is the expansion of RPC for a grid computing environment. An extended message passing interface (MPI) suitable to a grid environment (Grid-enabled MPI) enables to have a communication between heterogeneous distributed computers. Various types of grid-enabled MPI, STAMPI [4], MPICH-G [5], PACX MPI [6], and so on, have been developed. Researchers can have a cooperative execution by modifying their own simulation codes.

Although researchers can cooperatively execute various plasma simulation codes by using those technologies, those technologies have advantages and disadvantages. The cooperative execution of simulation codes are roughly divided into three types; type 1 is the sequential type in which the codes are executed sequentially, type 2 is the pipeline type in which simulation codes are executed concurrently by sending and receiving input/output data during their running, type 3 is the conditional branch type in which simulation codes are started and data is transferred depending on various conditions. In type 3, the choice and start of codes are judged during the execution of scientific scenario. Namely, some codes are started if a specified condition is satisfied at a conditional branch. Therefore, the schedule to execute all codes cannot be decided beforehand. A workflow tool is useful for type 1, but is not forte for type 2 and 3 since it does not always have a function for a detailed control such as conditional branch. On the other hand, RPC and Grid-enabled MPI are suitable for type 2 and 3. However, researchers have to make exertions to develop/extend a grid-enabled application and/or to modify their simulation codes.

The integration of plasma simulation codes, e.g. the "Burning Plasma Integrated Code", is classified in type 3, since their simulation codes have to be cooperatively executed depending on various conditions such as plasma stability, timing of heating, and so on. Therefore, it is desirable to construct the grid computing technology which enables researchers in nuclear field to control the type 3 execution without their exertions.

As a new framework to control various types of cooperative execution easily, we propose the Simple Orchestration Application Framework (SOAF) [7]. SOAF cooperatively controls simulation codes on remote computers from a client PC by only describing a configuration file which includes the information of execution codes and file flows among them. In addition, SOAF is designed to reduce modification of simulation codes and researchers' exertions for orchestration. SOAF has been developed by using client application program interfaces (APIs) implemented on Atomic Energy Grid Infrastructure (AEGIS) [8]. We have applied SOAF to the "Burning Plasma Integrated Code" which consists of various plasma simulation codes to solve the current diffusion, stability of plasma, current drive, and so on.

We describe SOAF in detail and its application to the "Burning Plasma Integrated Code" in Sects. 2 and 3, respectively. In Sect. 4, we report the environment and result of our experiment. Finally in Sect. 5, we summarize our R&D results and describe a future work.

2 Simple Orchestration Application Framework (SOAF)

2.1 Overview of SOAF

We propose a new framework which makes simulation codes cooperate without difficulty. Here we describe a framework concept that allows simulation codes to inter-operate. Here we mention a concept of our proposition.

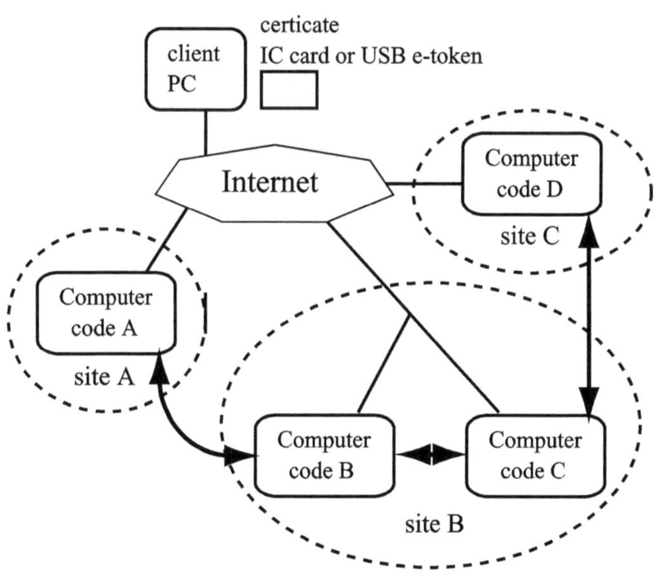

Fig. 1 Cooperative execution of simulation codes on distributed computers

We suppose that many simulation codes exist on distributed computers installed in different sites such as universities and institutes. This situation is shown in Fig. 1.

Each code can analyze one phenomenon in detail. When each phenomenon is not independent and affects each other, a realistic problem can be elucidated by cooperatively executing those simulation codes. To realize this, the orchestration including code executions and file transfers on a grid infrastructure is required. We have to consider file flow or flow of execution codes for orchestration. Here, we consider the followings are critical issue to design the SOAF:

1. How are various types of cooperative execution controlled easily?
2. How is each simulation executed cooperatively with its minimal modification?

We focused on a file flow among execution codes to design the SOAF. Firstly, it is better to send/receive information by transferring files in order to have communications among simulation codes without modifying those codes. Secondly, it is easier to define a file flow than a flow of execution codes in order to build a scientific scenario depending on conditional branch. A flow of execution codes cannot be described sequentially in case that a scientific scenario includes conditional branch. The start of codes needs to be managed by any way. We adopted the way to manage the start of codes by a file flow. It is useful to identify a file flow, since a code is usually started after it receives a file from another code, except a firstly started code. SOAF manages the start of codes and the transfer of files under a file flow which is described in a configuration file. SOAF consists of a client application, programs to support the file transfers, and a configuration file. Programs (we call this program "sentinel") are started by the client application (we call this "controller") and are executed on distributed computers. By those ideas, researchers in nuclear field can execute cooperatively various plasma simulation codes by only generating a configuration file.

The codes started by a conditional branch require input files. By directing our attention to the file flow among the codes, we can control the cooperative execution of the conditional branch type, because these files are generated by other codes.

2.2 Controller

We developed the controller using client APIs implemented on AEGIS. Center for computational science and e-systems (CCSE) of JAEA has developed AEGIS, which is grid middleware for atomic energy research. The schematic diagram of AEGIS is shown in Fig. 2. We have developed client APIs as a function of AEGIS to develop grid-enabled application on a client PC. Client APIs consist of low, middle, and high level APIs, which are classified due to their functions. To develop controller, we used the authentication API, file transfer API, job submission API and job information request API in low level APIs, and Job-script generator API in middle level APIs.

Low level APIs supply connection between client PC and supercomputers on AEGIS, job operation to supercomputers, resource handling on both clients PC and supercomputers, and so on. When users access AEGIS, users need an authentication process by using IC card or USB e-token (PKCS#11). Job-script generator API generates the script corresponding to heterogeneous computers by reading job attributes such as name of computer, job class, number of CPUs, path of program, work directory, and so on.

The client APIs are supplied as libraries which have an interface of C language, and are available for Linux and Windows.

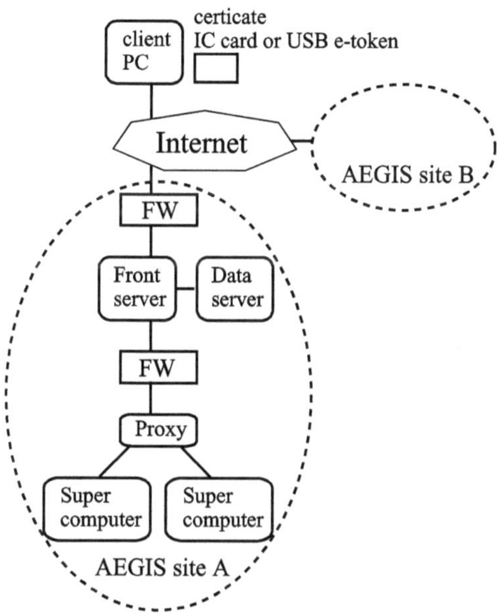

Fig. 2 Schematic diagram of AEGIS

2.3 *Sentinel*

The sentinel detects the output files and operates the file transfer between simulation codes. It works before and after job submission.

The sentinel consists of three scripts: **send**, **recv** and **void**. These scripts are described by Perl script language. The **send** script detects "flag files" of output files, generates "sent files". The **recv** script detects "sent files" and generates "flag files". The void script is used to execute simulation codes without files.

We show the procedure of file transfer using the sentinel. Here, we describe an example that two codes ("code A" and "code B") are cooperatively executed on distributed computers (see Fig. 3). The execution of "code B" requires output files from "code A". The procedure is as follows:

1. "Code A" generates output files and their "flag files".
2. The **send** sentinel beside "code A" detects the "flag files" and deletes them. Then **send** sentinel generates "sent files" of the output files. Finally **send** sentinel finishes (Before next file transfer, **send** sentinel is resubmitted again.).
3. By the detection of **send** sentinel beside "code A" end, the controller transfers the output files from work directory of "code A" to that of "code B".
4. The controller transfers "sent files" from work directory of "code A" to that of "code B".
5. The **recv** sentinel beside "code B" detects "sent files" and deletes them. Then the **recv** sentinel generates "flag files" of transferred files. Finally **recv** sentinel finishes (Before next file transfer, **recv** sentinel is resubmitted again.).
6. By the detection of **recv** sentinel beside "code B" end, the controller starts "Code B". "Code B" detects "flag files", deletes them and read the output files.

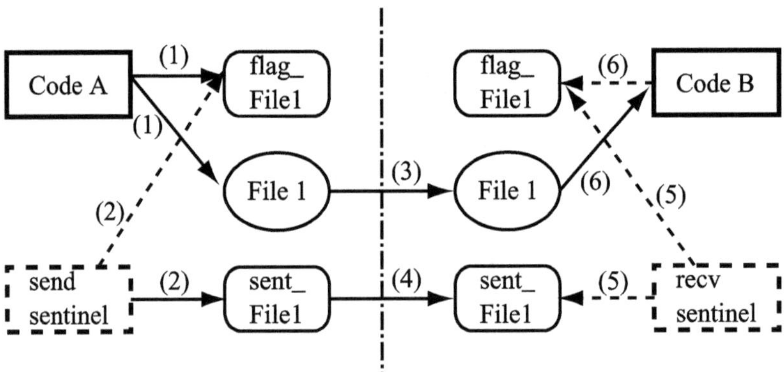

Fig. 3 The procedure of file transfer

A little modification of simulation codes is needed to use SOAF. When "code A" generates output files, "code A" must generate their "flag files". "Code B" must detect the "flag files" and delete them.

2.4 Configuration File

The configuration file in this case is described in Fig. 4. The configuration file consists of information of codes (PROGRAM) and file flow (FLOW). The "name" in PROGRAM is the name of the execution job. The "path" in PROGRAM means the path of execution code. The "type" in FLOW represents kinds of job. The "type 1" corresponds to the firstly started job. The job with "type 0" is started after it receives a file from another job. The "File1" is output files from code A. By the controller, these files are transferred from work directory of code A to that of code B. After files are transferred, the controller starts code B. The "File2" is output file from code B. After execution of code B, this file is transferred from work directory of code B to that of code A. Code A receives this file.

```
PROGRAMNUM 2
PROGRAM
  program     CodeA
  name        CodeA_0
  server      computer1.node.site
  path        /home/foo/bin/CodeA
  work        /home/foo/workdir
  queue       TSS
  para        1
  kind        serial
END

PROGRAM
  program     CodeB
  name        CodeB_0
  server      computer2.node.site
  path        /home/foo/bin/CodeB
  work        /home/foo/workdir
  queue       TSS
  para        1
  kind        serial
END
```

```
FLOWNUM 2
FLOW
  alias        CodeA_0
  exec         1
  type         1
  send         File1        CodeB_0
  recv         File2        CodeB_0
END

FLOW
  alias        CodeB_0
  exec         1
  type         0
  recv         File1        CodeB_0
  send         File2        CodeB_0
END
```

Fig. 4 An example of the configuration file

Because the controller is developed by using multithreaded process and each thread corresponds to each sentinel, the controller detects the end of file output and file transfer immediately. The execution time overhead of the sentinel is almost negligible.

3 Development of Simple Orchestration Application Framework

The previous GDS application lacked functionalities to realize the pipelined data-transfer scenario adapting two solutions proposed in Sect. 2.2 because the application was designed for the sequential data-transfer scenario. The application permitted only sequential control of data-transfer and job execution so that data could be transferred only after a job was completely terminated and there was no way to transfer the data during job's running. Besides, although the application regularly checked each job's status obtained from the AEGIS Client APIs, that was only for user's monitoring and when a job's abnormal end was informed, users needed re-submit the abnormal ended job themselves. To immediately resubmit a job, users must always keep their eyes on the monitoring messages.

To realize pipelined data-transfer scenario with these the solutions of "one time step execution" policy and job's immediate resubmit, we developed SOAF. In designing SOAF, we intended it to be general-purpose to also enable various other type scenarios necessary for other scientific simulations which couple multiple codes installed on distributed computing resources. For example, multi-physics or multi-scale simulations are composed of various scientific codes and each code is used with dedicated purpose for a specific physical phenomena or space-time scale. Coupling these codes involves data transfer between jobs in pipelined scenario, conditional branch scenario or even the mixed type of all these scenarios. It means that many scientific simulations with sophisticated coupling scenarios can be realized by controlling each scenario's data flow. Focusing on this point, we designed and developed SOAF and also considered that its implementation should be simple and use of its functionalities should be easy.

3.1 Burning Plasma Simulation

The "Burning Plasma Integrated Code" has been researched and developed mainly by Naka Fusion Institute of Japan Atomic Energy Agency (JAEA) to integratedly predict and interpret the plasma behavior. It consists of various plasma simulation codes to solve the current diffusion, stability of plasma, current drive and so on. A quite realistic simulation is expected by integrating those codes.

It is indispensable to understand the controllability of plasma toward the continuous operation of tokamak reactor especially for ITER [9]. To control the burning plasma and achieve high performance such as high confinement, high beta, high bootstrap, high radiation at the edge region, suppression of impurity, it has an important role to simulate behavior of burning plasma in tokamak reactor.

For simulation of burning plasma in tokamak reactor, it is not realistic to handle whole physical processes by one simulation code. One of the reasons is that burning plasma has very wide temporal and spatial ranges in the steady state (Fig. 5).

For example, a period of high frequency wave such as electron-cyclotron current drive is less than 10–8 seconds. On the other hand, current diffusion occurs in several seconds. Furthermore, the burning plasma has complex physics such as turbulence, transport, MHD, current diffusion, wave-particle interaction, plasma-wall interaction, atomic and molecular physics, and so on. Each physical process with different temporal and spatial scales is modeled and calculated by separated simulation codes. The respective simulation codes can be integrated for the burning plasma analysis. The integrated simulation system covers both microscopic and macroscopic physical processes and simulates long-time behavior considering short-time behavior.

Various plasma simulation codes for physical processes have been developed in universities and institutes. Since those codes are executed in each university and institute, we need a grid computing technology to cooperatively execute those codes.

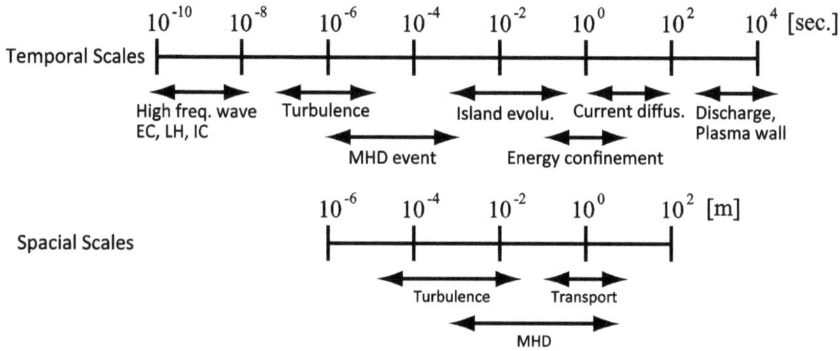

Fig. 5 Temporal and spatial scales on physical processes of plasma

Therefore, we have applied SOAF to manage those codes in the "Burning Plasma Integrated Code" on our grid infrastructure AEGIS. In the current application, we use four plasma simulation codes; tokamak prediction and interpretation code system (TOPICS) [10], two-dimensional magnetic stability analysis code (MARG2D) [11], electron-cyclotron current drive code (ECCD) [12], and lower-hybrid current drive code (LHCD).

TOPICS solves the 1D transport and current diffusion equations and the Grad-Shafranov equation of the MHD equilibrium on the 2D plane. The transport code solves the current diffusion equation, including EC and LH current profiles. TOPICS investigates specific characteristics of burning plasma such as behavior of edge localized modes (ELMs) [13] and neoclassical tearing modes (NTM) stability.

MARG2D is 2-D Newcomb equation solver which solves an eigenvalue problem associated with the two-dimensional Newcomb equation in axisymmetric toroidal plasma such as tokamak by using a finite element method. Using this code, we can analyze stability of ideal external MHD modes from low to high toroidal mode numbers. Furthermore we obtain eigenfunctions numerically which show the singular behavior. Using MARG2D, the MHD property of JT-60SA, the complemented device of ITER, is investigated [14].

ECCD and LHCD codes simulate control and stabilization of the burning plasma and thus are executed for control of burning plasma simulated by TOPICS.

MARG2D, ECCD, and LHCD are started depending on the requirement arising from TOPICS during the burning plasma simulation. When the instability modes are found the plasma is found to be close to the unstable region by MARG2D, TOPICS requests to start ECCD or LHCD for stabilization of the burning plasma. Therefore the "Burning Plasma Integrated Code" belongs to the conditional branch type.

Table 1 Codes and computers

Codes	Computers
TOPICS	Altix350
MARG2D	Altix3700Bx2
LHCD	Altix350
ECCD	PC Cluster

Table 2 The situation of the simulations

Simulation time t (sec)	Action
0	TOPICS start
1	MARG2D execute
2	LHCD execute
4	ECCD execute
5	TOPICS finish

4 Experiment

In this section, we mention our experiment about the application of SOAF to those four codes in "Burning Plasma Integrated Code".

The client PC where the controller is executed is Ubuntu Linux 8.04 located in CCSE/JAEA (Tokyo/Japan). The C compiler is gcc-4.2.4 (multithread enabled). We used USB e-token (PKCS#11) for authentication to AEGIS.

The numerical simulations are submitted to three computers located in Tokai Research and Development Center of JAEA (Ibaraki/Japan); Altix3700Bx2, Al-tix350, and PC cluster. In this experiment, we fix the computers in which simulation codes are executed as shown in Table 1. In this experiment, TOPICS, LHCD, and ECCD are executed on TSS mode (serial execution). MARG2D is submitted to job queuing system (class of 32CPUs and 3 hours). When we start the controller, not only TOPICS but also each sentinel script beside simulation code is executed. All sentinel scripts are executed on TSS mode.

Those simulation codes are executed by the following procedure. At first, TOPICS is started. In this experiment, the simulation time (not CPU time) is set to 5 seconds ($t = 5$). During the running of TOPICS, TOPICS requests to start MARG2D, LHCD, and ECCD. In this experiment, the start of codes is scheduled as shown in Table 2. TOPICS generates output files for MARG2D at simulation time $t = 1$. Then these output files are transferred to work directory of MARG2D. After the file transfer, MARG2D is started. LHCD and ECCD are started at simulation time $t = 2$ and 4, respectively. In the configuration file of controller, TOPICS corresponds to "type 1" code. MARG2D, LHCD, and ECCD correspond to "type 0" codes.

The diagram of the simulations is shown in Fig. 6. During simulation by MARG2D, LHCD, and ECCD, TOPICS suspends. Then TOPICS reflects the analysis results by other codes and restarts.

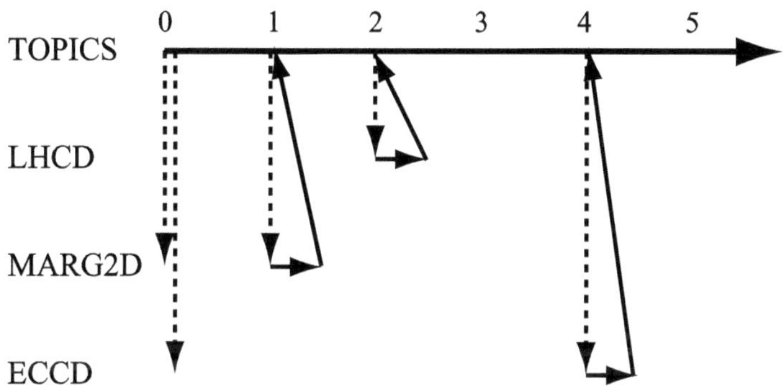

Fig. 6 The diagram of the simulations. The solid line and dashed line mean code execution and file transfer, respectively

The file flow is shown in Fig. 7. The files are transferred between TOPICS and other 3 codes. MARG2D, LHCD, and ECCD receive the input files from TOPICS and return results of analysis.

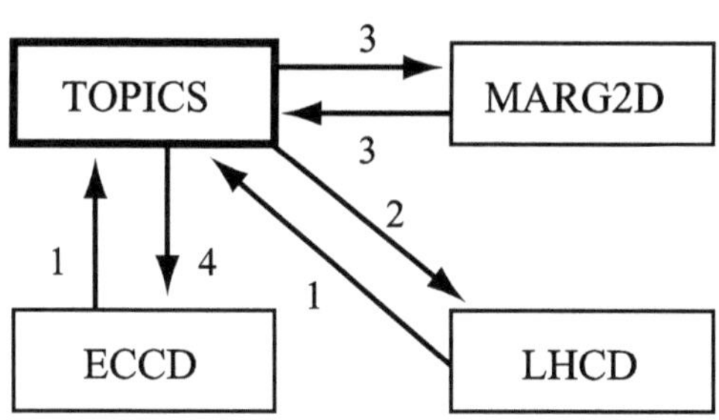

Fig. 7 The file flow between simulation codes. The number beside arrow means the number of file

TOPICS generates three files with flag files just after the starting: two files for MARG2D and one file for ECCD. These files are transferred by the controller which detects the end **send** script beside TOPICS immediately. After transfer, **send** script beside TOPICS restarts.

At $t = 1$, TOPICS generates a file and its "flag file" for MARG2D. During analysis of MARG2D, TOPICS suspends. The **send** script beside TOPICS detects "flag file" from TOPICS and deletes it. Then the **send** script generates "sent file" and ends. The controller detects the end of **send** script beside TOPICS and transfers

Table 3 The situation of the simulations

Wall-clock time (min)	Action
0	SOAF start TOPICS start
4	MARG2D start
5	MARG2D finish
25	LHCD start
26	LHCD finish
33	ECCD start
36	ECCD finish
0	TOPICS finish SOAF finish

one output file and a "sent file" to work directory of MARG2D. The **recv** script beside MARG2D detects "sent files" from TOPICS and ends. Then the controller detects the end of the **recv** script and submits a job of MARG2D. After execution of MARG2D, the **send** script detects "flag files" of 3 output files and deletes them. Then the **send** script generates "sent file" and ends. The controller detects the end of **send** script beside MARG2D and transfers output files and "sent files" to work directory of TOPICS. The **recv** script beside TOPICS detects these "sent files" and generates "flag files". Then the **recv** script ends. Finally TOPICS detects "flag files" and restarts.

We prepare the configuration file for four simulation codes to execute the controller. We have confirmed that the controller executes those simulation codes as scheduled at Table 2. SOAF successfully controls the cooperative execution of four simulation codes and file transfers between those codes.

The performance of our experiment is shown in Table 3. The whole execution time is about 40 minutes without job queuing time of MARG2D. Because the size of transferred files is less than 1 Mbytes, each file transfer is completed less than 1 minute. In this experiment, the controller generates 18 threads for the sentinels: 1 for TOPICS execution (void script), 14 for I/O on TOPICS, 3 for each simulation codes (MARG2D, ECCD, LHCD). The overhead of the controller is about 2 minutes and does not almost influence to the total execution time. Because the most of this execution time is used for the authentication to AEGIS, the overhead is not considered to increase substantially even if a cooperative simulation becomes complicated.

The timing of start for ECCD and LHCD is scheduled by the conditional branch implemented in TOPICS. Because the start of ECCD and LHCD is decided by only the existence of output files from TOPICS, SOAF is applicable for any conditional branch.

By using SOAF, we can execute cooperatively various simulation codes on remote computers from a client PC without difficulty. We can achieve the orchestration of simulations with minimal modification. The length of "Burning Plasma Integrated Code" is about 300,000 lines. The modification for cooperative execution is only about 200 lines. If we apply GridRPC or Grid-enabled MPI, the modification would become several thousand lines. We have verified that all of our issues are solved.

5 Summaries

We have developed the Simple Orchestration Application Framework (SOAF) to cooperatively control simulation codes on remote computers from a client PC on our grid infrastructure AEGIS. Researchers can easily execute cooperative simulation codes by using a configuration file in which a file flow is described. Another advantage of SOAF is that researchers can cooperatively control simulations without substantial modification of simulation codes. And, the indication of cooperation can be easily changed by only rewriting a configuration file.

We confirm the usefulness of SOAF by applying it to the "Burning Plasma Integrated Code". In the current experiment, we use four simulation codes on distributed computers. SOAF can control cooperative executions of these four simulation codes and file transfers between them.

We have solved two issues described in Sect. 2. Although we show an example for the conditional branch type, researcher can integrate various kinds of simulation codes by just describing the data dependency among these codes. Therefore, the first issue is solved. In order to notify the timing of file transfer to SOAF, only a few SOAF library calls are inserted in the applications. The second issue is solved.

From our experiment, we verified further merit of SOAF. For cooperative execution of simulation codes, the configuration file for the controller is required. The length of the configuration file is much less than that of the simulation codes. The overhead for SOAF is much less than that for simulations.

For consideration of the realistic situation, we need longer-time simulation. In the current experiment, the simulation time of tokamak plasma is only 5 seconds. For example, the burn duration in ITER project is designed as more than 1000 seconds [9]. When burning plasma in such a situation is simulated, we would require several weeks or several months. To achieve this, we need to consider the continuity of execution. The longer-time simulation may suffer from various unexpected stop which is caused by execution time excess, queuing timeout, outage of computers and so on. To avoid the stop of the simulation, we will apply the fault-tolerant mechanism of SOAF to the "Burning Plasma Integrated Code". When this mechanism is applied, a long-time simulation can be executed automatically.

Acknowledgements We would like to thank Dr. T. Ozeki, Dr. N. Hayashi, and Mr. K. Iba for fruitful discussion and support and Mr. I. Kamata for modification of the simulation codes.

References

1. Ludäscher, B., Altinta, I., Berkley, C., Higgins, D., Jaeger, E., Jone, M., Lee, E.A., Tao, J. and Zhao, Y. (2005) Scientific workflow management and the Kepler system. Concurrency and Computation: Practice and Experience 18:1039–1065.
2. Imamura, T., Hasegawa, Y., Yamagishi, N. and Takemiya, H. (2002) TME: A Distributed resource handling tool. In Recent Advances in Computational Science & Engineering, International Conference on Scientific & Engineering Computation (IC-SEC) (3–5 December 2002, Raffles City Convention Centre, Singapore):789–792.
3. Seymour, K., Nakada, H., Matsuoka, S., Dongarra, J., Lee, C. and Casanova, H. (2002) Overview of GridRPC: A Remote Procedure Call API for Grid Computing. In: Parashar, M. (ed). Proceedings of 3rd International Workshop on Grid Computing:274–278.
4. Imamura, T., Tsujita, Y., Koide, H., and Takemiya, H. (2000) An Architecture of Stampi: MPI Library on a Cluster of Parallel Computers. Proceedings of the 7th European PVM/MPI Users' Group Meeting on Recent Advances in Parallel Virtual Machine and Message Passing Interface:200–207.
5. Foster, I. and Karonis, N.T. (1998) A grid-enabled MPI: message passing in heterogeneous distributed computing systems. Proceedings of the 1998 ACM/IEEE conference on Supercomputing (CDROM):1–11.
6. Gabriel, E., Resch, M., Beisel, T. and Keller, R. (1998) Distributed Computing in Heterogeneous Computing Environment. Proceedings of the 5th European PVM/MPI Users' Group Meeting on Recent Advances in Parallel Virtual Machine and Message Passing Interface:180–187.
7. Suzuki, Y., Nakajima, K., Kushida, N., Kino, C., Minami, T., Matsumoto, N., Aoyagi, T., Nakajima, N., Iba, K., Hayashi, N., Ozeki, T., Totsuka, T., Nakanishi, H. and Nagayama, Y. (2008) Research and development of fusion grid infrastructure based on atomic energy grid infrastructure (AEGIS). Sixth IAEA Technical Meeting on Control, Data Acquisition, and Remote Participation for Fusion Research (4–8 June 2007, Inuyama, Japan), Fusion Engineering and Design 83:511–515.
8. Tatekawa, T., Nakajima, K., Kim, G., Teshima, N., Suzuki, Y., Takemiya, H., Hayashi, N. and Iba, K. (2010) Simple Orchestration Application Framework to control "Burning Plasma Integrated Code", Proceedings of The Third International Joint Conference on Computational Sciences and Optimization (CSO2010) 2:322–326.
9. ITER Physics Basis Editors et al. (1999) Chapter 1: Overview and summary. Nucl. Fusion 39:2137–2174.
10. Hayashi, N., Isayama, A., Nagasaki, K. and Ozeki, T. (2004) Numerical Analysis of Neoclassical Tearing Mode Stabilization by Electron Cyclotron Current Drive. J. Plasma Fusion Res. 80:605–613.
11. Tokuda, S. and Watanabe, T. (1999) A new eigenvalue problem associated with the two-dimensional Newcomb equation without continuous spectra. Phys. of Plasmas 6:3012–3026.
12. Hamamatsu, K. (1999) Numerical Study for Positional Control of ECCD by the Ordinary Wave in a Tokamak Plasma. J. Plasma Fusion Res. 75:143–150.
13. Hayashi, N., Takizawa, T., Ozeki, T., Aiba, N. and Oyama, N. (2007) Integrated Simulation of ELM Energy Loss Determined by Pedestal MHD and SOL Transport. Nucl. Fusion 47:682–688.
14. Aiba, N., Tokuda, S., Fujita, T., Ozeki, T., Chu, M.S., Snyder, P.B. and Wilson, H.R. (2007) Numerical Method for the Stability Analysis of Ideal MHD Modes with a Wide Range of Toroidal Mode Numbers in Tokamaks. Plasma Fusion Res. 2:010.

Part IV
Acoustics and Structural Mechanic

On Sound Generated by a Globally Unstable Round Jet

G. Geiser, H. Foysi, W. Schröder, M. Meinke

Abstract Direct numerical (DNS) and large-eddy simulations (LES) of a strongly heated globally unstable round jet are juxtaposed with respect to aerodynamical mean characteristics and the sound being generated. The sound field is computed by a hybrid approach using the acoustic perturbation equations (APE). All used codes have been adopted to massive-parallel supercomputers. This way results can be obtained in a reasonable time frame. When compared to the DNS results, the LES is capable to capture the major characteristics of the emitted sound field in the forward direction. The sideline and backward direction that are dominated by small scale noise reveal larger discrepancies that are due to the inherent restrictions of large eddy simulations.

1 Introduction

The noise of heated jets has been investigated intensively in the past. However, there still are many open questions which require further research. For example, the mechanism which produces the change in spectral shape at large aft angles is still unknown. Additionally, the precise effect of the jet density on the radiated noise needs further investigations. Recently, Viswanathan [40] undertook a systematic experimental study of heated jets and generated a database for low jet velocities ($U_j/a_\infty \leq 0.6$). He could confirm the existence of an extra sound source of dipole type at high temperatures to be due to Reynolds number effects. However, measurements at such small velocities are difficult, since the noise amplitude generated by valves, ducts, rigs, etc. may have larger magnitudes than the jet itself.

Alternatively, numerical simulations can be performed. Unfortunately, direct numerical simulations (DNS) at Reynolds numbers which are high enough (> 400000)

G. Geiser, H. Foysi, W. Schröder, M. Meinke
Institute of Aerodynamics, RWTH Aachen University, Wüllnerstraße 5a, 52062 Aachen, Germany
e-mail: G.Geiser@aia.rwth-aachen.de

M. Resch et al. (eds.), *High Performance Computing on Vector Systems 2010*,
DOI 10.1007/978-3-642-11851-7_9,

to avoid low Reynolds number effects are unrealistic in the near future. DNS or large-eddy simulations (LES) of turbulent flows have shown promising results in predicting the noise of high Reynolds number jets [42, 4, 5, 2, 3, 6, 22, 19, 25]. For many applications, a direct calculation of the far-field is still computationally too expensive. Therefore, hybrid methods are being used, separating flow and sound calculation [18, 13, 19, 25]. The quality of the calculated noise certainly depends on the acoustic model, grids, and numerical method being used. Furthermore, the effect of the unresolved scales in LES calculations on the emitted sound spectrum needs to be quantified.

The present paper investigates the sound generated by a strongly heated globally unstable round jet using both DNS and LES for flow simulation. The acoustic field is determined in a second step by solving the acoustic perturbation equations [13].

The paper is organized as follows. First, a description of the numerical methods and the parallelization of the acoustic solver is given. Then, characteristics of the round jet DNS and LES and the acoustic fields obtained by the hybrid calculations are being presented before a short summary finalizes the paper.

2 Numerical Setup

2.1 Round Jet Flow Simulation

The Navier-Stokes equations for compressible flow are solved in cylindrical (x,Φ,r) coordinates, together with a transport equation for a passive scalar field ξ. The centerline singularity at $r = 0$ is treated by applying the method of Mohseni & Colonius [32]. For time integration a low-dispersion-low-dissipation fourth-order Runge-Kutta scheme of Hu *et al.* [21] in its low storage form is used. The spatial differentiation utilizes optimized explicit DRP-SBP (dispersion-relation-preserving summation by parts) finite-difference operators of sixth order [23]. For the LES simulations, the compressible form of the dynamic Smagorinsky model [30] is applied. For numerical stability the simulations require the timestep Δt to be such that the CFL number is restricted by $CFL < 1.4$. Time accuracy of the solution restricts the CFL number even further to $CFL \leq 1.1$ [33]. This numerical scheme yields numerical errors which are small compared to the subgrid-scale terms of the governing equations when the filter size is chosen twice the grid size [11], as it is done here. Table 1 summarizes the different simulation parameters.

The initial momentum flux m_j has been fixed for all simulations and was chosen similarly as in Wang *et al.* [41]. This reasoning is based on observations of Ricou & Spalding [34] who showed that the entrainment rate of a jet strongly depends on m_j. Ruffin *et al.* [35] demonstrated additionally, that the Taylor and Kolmogorov length scales are determined by the initial momentum flux which was also confirmed by other experiments and simulations [1, 12, 41]. Non-reflecting boundary conditions [29] were applied, together with a combination of grid stretching and

Table 1 Parameters of the heated round jet simulations. The Reynolds number is given by $Re = U_j \rho_j D/\mu_j$. $\delta_{\theta 0}/D$ denotes the initial momentum thickness normalized by the jet diameter D. The density ratio of the jet is defined by $s = \rho_j/\rho_{co}$. The domain lengths L_i were normalized by r_j. The number of grid points in the respective coordinate directions are represented by n_i

Case	Re	$D/\delta_{\theta 0}$	s	L_x	L_Φ	L_r	n_x	n_Φ	n_r
DNS014	7000	54	0.14	50	2π	14	1024	256	320
LES014	7000	27	0.14	60	2π	16	256	64	112

spatial filtering close to the boundaries to damp any disturbances reflected from the boundaries. A constant grid spacing was chosen within the range $0 < r/r_j < 1.5$ and $0 < x/r_j < 12$, where r_j denotes the initial jet radius. A subsequent relative stretching of 1% was used extending in the r-direction up to $r/r_j = 10$ and $r/r_j = 12$ in the radial and up to $x/r_j = 40$ and $x/r_j = 50$ in the streamwise direction for case DNS014 and LES014, respectively. In the remaining part of the domain the stretching ratio was increased to 6%. For the LES grid, the number of grid points within $0 \leq r/r_j \leq 1$ was 11. The LES was run using 32 processors on the JUMP supercomputer of the Jülich Supercomputing Centre. The computation including statistics required a CPU time of approximately 24 h. The DNS simulations were run on 4096 processors on a Blue-Gene/P system at the same location. Hybrid shared-/distributed-memory parallelization was applied to enhance the parallel performance. The round jet simulations were initialized using velocity profiles of the form [2]

$$U = U_{co} + \frac{1}{2}(U_j - U_{co})\left(1 - \tanh\left[\left(\frac{r}{r_j} - \frac{r_j}{r}\right)\frac{1}{4\delta_{\theta 0}}\right]\right),$$

and temperature fields using the Crocco-Busemann relation [9]. The passive scalar field was prescribed initially using a tanh function [17]. The initial normalized momentum thickness in the above profiles was prescribed according to Table 1. To trigger the transition to turbulence precursor simulations of an annular temporally evolving mixing layer with smaller initial momentum thickness were performed. The square root of the turbulent kinetic energy, non-dimensionalized using the mean values at the jet inlet, was monitored until its value $\sqrt{(U')_1^2 + (U')_2^2 + (U')_3^2}$ was of the order of 0.05 and the mean velocity profiles agreed closely with the prescribed inlet mean jet profile. The resulting fluctuations were convected into the fluid domain using the characteristic inflow boundary conditions. Additionally, solenoidal broadband velocity fluctuations [36] and stochastic vortex ring perturbations [5] were superimposed on the annular mixing layer to break the periodicity of the inflow data. Although the domain sizes and number of grid points are similar to those chosen by Bodony *et al.* [2], the grid resolution of the LES case was checked by performing also higher resolved simulations ($n_x \times n_\theta \times n_r = 384 \times 80 \times 120$) [16]. These simulations showed negligible differences regarding the mean flow statistics presented in this paper. The DNS simulation has a grid spacing approximately 6 times the Kolmogorov scale η, which was sufficient for the calculation of the acoustic far field [24]. Simulations at higher Reynolds number and different density ratios are conducted presently, having a grid spacing of 2.5η to 4η.

The statistical quantities were obtained by averaging in the homogeneous directions and over a time period T, corresponding to a Strouhal number $D/(TU_j) = 9 \cdot 10^{-4}$, after statistical steadiness was reached.

Field dumps were obtained from the DNS and LES at a sampling frequency of $St_D = 16$ and $St_D = 20$, respectively. From these dumps the acoustic sources were computed in a separate preprocessing step.

2.2 Aeroacoustic Computation

The aeroacoustic computations are being performed by solving the acoustic perturbation equations (APE) [13]. This set of equations was derived from the viscous conservation laws by applying source filtering based on an eigendecomposition in Fourier/Laplace space. Only transportation effects related to acoustical eigenmodes contribute to the operator on the left-hand side, while the remaining terms form the corresponding acoustical sources on the right-hand side. The APE offer stable linear acoustical propagation in arbitrary mean flows while taking into account convection and refraction effects. They have been successfully applied to several aeroacoustics problems including trailing edge noise [14], high-lift airfoil noise [26], combustion noise [8, 7], and jet noise [25, 19].

Since compressible fluids are considered, the APE are used in their APE-4 formulation

$$\frac{\partial p^a}{\partial t} + \bar{c}^2 \nabla \cdot \left(\bar{\rho} \boldsymbol{u}^a + \bar{\boldsymbol{u}} \frac{p^a}{\bar{c}^2} \right) = \bar{c}^2 q_c \tag{1}$$

$$\frac{\partial \boldsymbol{u}^a}{\partial t} + \nabla \left(\bar{\boldsymbol{u}} \cdot \boldsymbol{u}^a \right) + \nabla \left(\frac{p^a}{\bar{\rho}} \right) = \boldsymbol{q_m}, \tag{2}$$

where ^{a} denotes acoustical perturbations that are the unknowns of the present system of equations and q_c and $\boldsymbol{q_m}$ are the source terms given by

$$q_c = -\nabla \cdot (\rho' \boldsymbol{u}')' + \frac{\bar{\rho}}{c_p} \frac{\overline{\mathrm{D}} s'}{\mathrm{D}t} \tag{3}$$

$$\boldsymbol{q_m} = -(\boldsymbol{\omega} \times \boldsymbol{u})' + T' \nabla \bar{s} - s' \nabla \bar{T} - \left(\nabla \frac{(\boldsymbol{u}')^2}{2} \right)' + \left(\frac{\nabla \cdot \tau}{\rho} \right)'. \tag{4}$$

$\overline{(\ldots)}$ indicates Reynolds averaged mean and $(\ldots)' = (\ldots) - \overline{(\ldots)}$ perturbation quantities. Due to the high Reynolds number viscous sound sources can be neglected for the cases investigated.

The APE are discretized by an alternating 5–6 stage low-dispersion low-dissipation Runge-Kutta (LDDRK) scheme [21] while the space derivatives are determined explicitly using a 7-point 4th-order dispersion-relation preserving (DRP) scheme [37]. On the boundaries non-reflecting radiation boundary conditions [37] are imposed. Non-physical high frequency waves are suppressed by artificial selective damping [38].

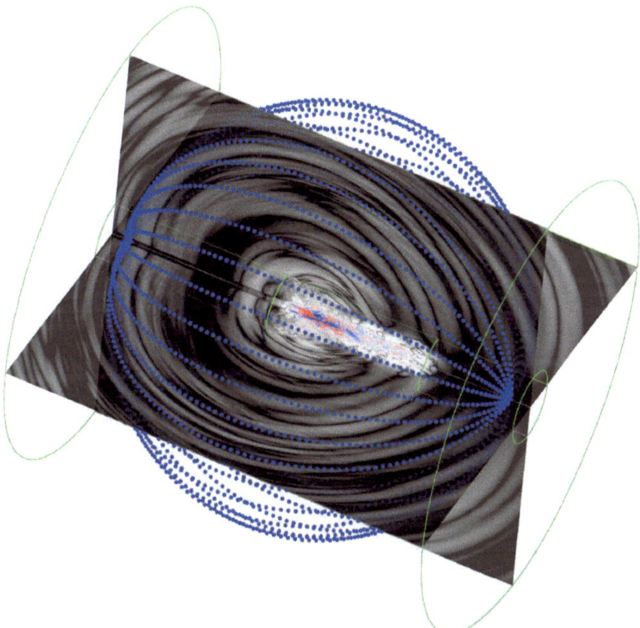

Fig. 1 Acoustic domain equipped with virtual microphones given as blue spheres. Green lines indicate the edges of the sub-domains. The slices show an instantaneous snapshot of the acoustic perturbation pressure p^a

The acoustic domains have a streamwise, radial, and azimuthal extension of $x \approx -22.5D \ldots 32.5D$, $r \approx 0D \ldots 27.5D$, and $\Phi = 0 \ldots 2\pi$, respectively. The maximum grid spacing in streamwise and radial direction is 0.2 and 0.26 for the DNS and LES case, respectively. The domains are equipped with virtual microphones measuring p^a, c.f. Fig. 1. These microphones are arranged on circles with $\Delta\Theta = 2°$. A sphere with radius $r = 25D$ is formed by 18 circles resulting in $\Delta\Phi = 10°$. The sphere is centered at $(x, y, z) = (5D, 0D, 0D)$ which approximately corresponds to the end of the jet's potential core. During postprocessing the overall sound pressure level (OASPL) and power spectral density (PSD) estimates on the spheres are averaged onto a half circle, since the sound generation is assumed to be statistically isotropic in the azimuthal direction. Power spectral density (PSD) estimates are computed using Welch's method with overlapping Blackman-Harris windows.

The acoustic computations were performed by using 2048–4096 cores of the massively parallel IBM BlueGene/P supercomputer JUGENE at the Jülich Supercomputing Centre. Wallclock time was approximately 40 h for the DNS and 4 h for the LES case, respectively.

2.3 Parallelization of the Acoustic Solver

The acoustic field is computed with the structured multi-block solver PIANO that uses curvilinear meshes for spacial discretization. PIANO is developed by the German Aerospace Center (DLR) in cooperation with the Institute of Aerodynamics. The code was adopted for computations on massive-parallel supercomputers by using non-blocking communication via the Message Passing Interface (MPI) [31]. The exchange of three ghostlayers is required for each subdomain. Communication bandwidth has been enhanced by collocating data to single messages when possible.

The multi-block structured domain is partitioned into the required amount of structured sub-domains by a weighted tree based cutting technique. Each block is cut preferentially orthogonal to its largest dimension to minimize the growth of internal surfaces, while the smaller dimensions are cut with a lower priority. The positions of the specific cuts are determined with respect to balanced sizes of the resulting blocks. No additional points are introduced by the partitioning, i.e. adjacent blocks have no common data on the generated cuts. The load is balanced with respect to the total number of grid points. Each core determines its required send and receive operations by exchanging requests with all cores in a preprocessing step. Domain partitioning, load balancing, and communication preparation is performed as needed by the acoustic solver and allows computations independent of the requested number of cores, only limited by the smallest size allowed for the sub-domains and the memory available to each core.

Figure 2(a) plots the parallel speedup of the PIANO code on a massive-parallel system. The speedup has been normalized to 128 cores, since this is the smallest

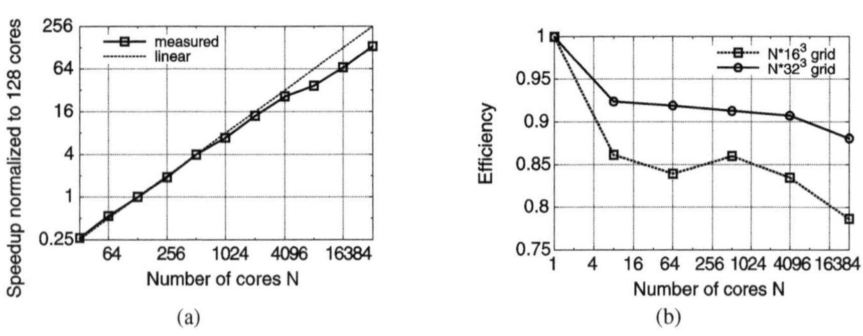

Fig. 2 (a) Parallel speedup of the acoustic solver PIANO on IBM BlueGene/P using a 256^3 grid with periodic boundary conditions. The speedup value is normalized to 128 cores. (b) Parallel efficiency of the acoustic solver PIANO on IBM BlueGene/P using a $N \times 16^3$ and $N \times 32^3$ grid with periodic boundary conditions. N denotes the number of cores being used

amount of cores that is allocatable on the used machine. For lower amounts of cores some cores are idle that probably causes higher speedup results. Slight variations from linear scaling can be observed for 256 cores that is due to the impossibility to equally partition the domain. At 512 cores a equal partitioning can be found that results in a linear speedup. At higher numbers of cores inter-rack communication is required that is subject to higher latency and lower bandwidth. This probably causes the reduced speedup, though retaining linearity again. Current revisions of the MPI standard lack methods to optimally map the topology of the computational domain to the topology of the supercomputer. These methods will be part of future MPI revisions and will probably increase the performance for general multi-rack computations. Furthermore the volume-to-surface ratio for the sub-domains becomes really small for large numbers of cores, which causes a high percentage of work related to communication.

Figure 2(b) depicts the parallel efficiency of the PIANO code for a $N \times 16^3$ and $N \times 32^3$ grid, respectively, where N denotes the number of cores used. The parallel code is in general slower when compared to the serial one. This is caused by the need for MPI function calls and additional copy operations to and from the communication buffers, while the serial code directly puts data to its final memory location. The parallel efficiency only slightly decreases with a raising number of cores, except for the largest number of cores tested that requires heavy inter-rack communication. The 32^3 case performs better which is caused by its better volume-to-surface ratio, resulting in a relatively lower portion of communication in the overall workload. However, this case turned out to be more comparable with the jet noise computations presented in this work. Therefore a reasonable parallel performance was achieved. Further improvements in parallel performance could be achieved by asynchronous numerical operations to be performed while communications takes place and the usage of shared-memory parallelization where possible.

I/O performance turned out to be the bottleneck of the present hybrid CFD/CAA approach on massively parallel supercomputers. For the acoustic sources, data in the order of Terabytes has to be read throughout the entire simulation and results of the in general large acoustic domain have to be written for analysis and visualization purposes. Furthermore the data should not be subject to domain partitioning. In this case the file layout would depend upon the number of cores used for the computation and the computational domain is scattered into many tiny sub-domains, which complicates pre- and post-processing. To ease handling of data and guarantee its reusability in the future, usage of a standardized data format is highly recommended. The most matured format at present time is HDF5 [39], since it offers a flexible structure supporting large datasets. Furthermore parallel I/O and partial data access via hyperslabs is supported. PIANO utilizes HDF5's collective parallel and partial file access to keep data in its original layout inside the files, while being independent of the number of cores involved in the computation. HDF5 showed usable performance, but, however, it does not scale well to a large number of cores. Furthermore HDF5 lacks support of asynchronous I/O. Of course this could be achieved by creating distinct I/O threads, though there is no reliable threading support for Fortran, that is still probably the most widely used programming language in computational fluid dynamics. The parallel access to unstructured data in HDF5 is also not yet sufficient. Improvement of these drawbacks would be highly advantageous for future massive-parallel computational fluid dynamics (CFD) and computational aeroacoustics (CAA) applications.

3 Results

3.1 Jet Characteristics

The decay of the centerline velocity U_c of variable density jets has been investigated by many researchers. To evaluate the quality of the present simulations, the findings were compared to data by Wang *et al.* [41] and the experiments by Amielh *et al.* [1]. Details for the round jet cases may be found in Foysi *et al.* [16]. As an example, the centerline velocity decay of the round jet LES case, along with data of Wang *et al.* [41] and the similarity law proposed by Chen & Rodi [10]

$$U_c/U_j = 6.3(\rho_j/\rho_{co})^{1/2}(2r_j/x), \tag{5}$$

is shown in Fig. 3(a). Although Wang *et al.* [41] calculated confined jets using fully developed pipe flow as inlet condition, the agreement is very good. As a further check, the normalized radial distribution of the streamwise velocity fluctuations at various streamwise locations versus the radial distance are depicted in Fig. 3(b), for the same simulation. They show fair agreement and a collapse of the data is observed, when normalized using the centerline velocity and the jet half-width.

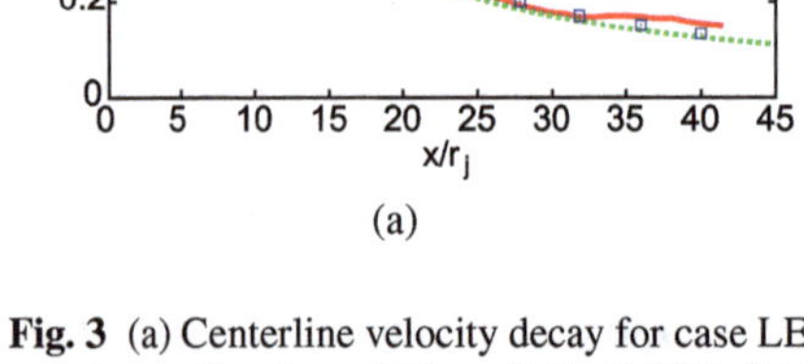

(a)

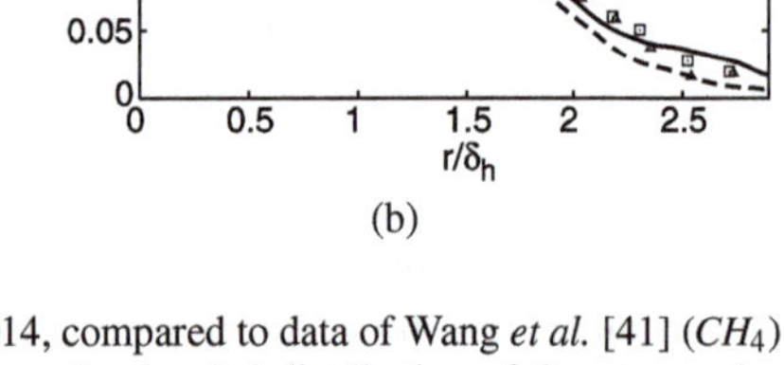

(b)

Fig. 3 (a) Centerline velocity decay for case LES014, compared to data of Wang *et al.* [41] (CH_4) and the scaling law of Chen & Rodi [10]. (b) Normalized radial distribution of the streamwise velocity fluctuations at various streamwise locations versus the radial distance, compared to data of Wang *et al.* [41]

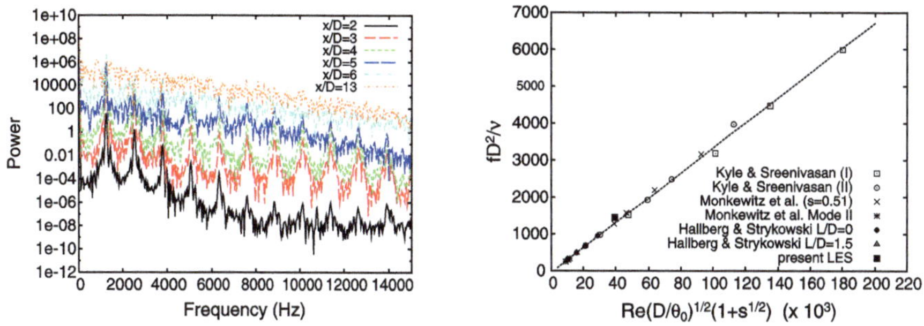

(a) (b)

Fig. 4 (a) Frequency power spectrum of the streamwise velocity at various streamwise positions (curves shifted for better visibility). (b) Roshko number fD^2/ν of the global instability mode, compared with the universal scaling of Hallberg & Strykowski [20] (L/D denotes the length of the different extension tubes used in the experiments)

In the LES and DNS cases at a density ratio of $s = 0.14$, high amplitude oscillations were observed, especially for the streamwise velocity and density, leading to strong gradients which made the simulation unstable. Adding artificial diffusivity [15] stabilized the calculation. This strong oscillating mode together with the strong vortex pairing events as well as the occurrence of side jet phenomena points toward the existence of a global instability in this simulation. Evidence for this is shown in Fig. 4(a), where the power spectrum of the streamwise velocity at various streamwise locations is plotted. The power spectrum clearly shows the fundamental frequency and its sub-harmonics at various streamwise positions (shifted for better visibility). The peak is visible until a sudden transition to turbulence has occurred around $x/D = 7$. The Strouhal number of this oscillating mode $St_D = fD/U_j$ was calculated and compared to various experimental datasets, depicted in Fig. 4(b). The data has been scaled following the proposal of Hallberg & Strykowski [20], who showed that the global oscillations depend on the density ratio s, $D/\delta_{\theta 0}$, and the Reynolds number, as long as compressibility and buoyancy effects are unimportant. Going further than Kyle & Sreenivasan [27], Hallberg & Strykowski [20] non-dimensionalized the frequency by the *viscous* time scale D^2/ν, thus retaining the Reynolds number in the frequency dependence. Therefore, they were able to collapse all data onto a straight line, when plotting the Roshko number fD^2/ν over $Re\sqrt{D/\delta_{\theta 0}}(1+\sqrt{s})$. Excellent agreement is found and the data almost collapses onto a single line. This strong oscillation and intense potential core collapse indicates a profound influence of the global instability on the radiated sound spectrum, since it is well known that there is a close link between turbulence in the core region and the sound generation [28]. Figure 5 illustrates the strong collapse, by showing contours of the passive scalar field ξ, which indicates this sudden breakdown.

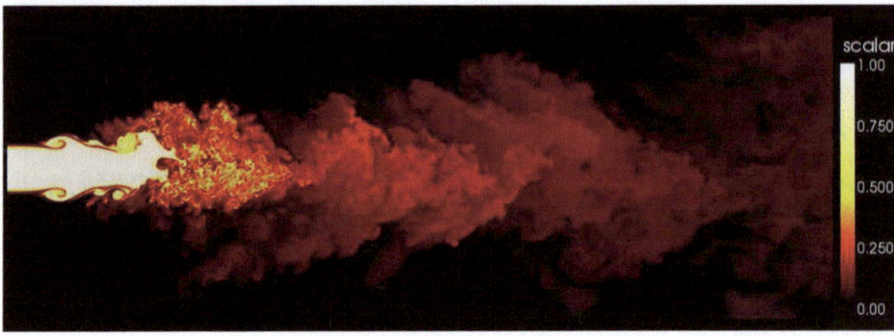

Fig. 5 Visualization of the contours of the passive scalar field ξ in simulation DNS014

3.2 Acoustic Results

The LES contains a burst phenomenon that is probably caused by a so-called side jet. These random events can occur for jets at that low density ratios s. This burst turned out to dominate the sound being propagated to the far-field. To evaluate the jet's regular behavior two different time-frames were chosen according to Fig. 6, where most of the burst phenomenon is excluded in the LES A case.

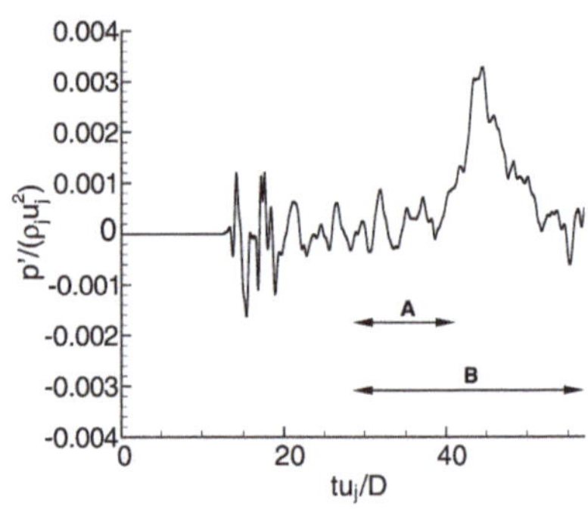

Fig. 6 Acoustic perturbation pressure measured by the virtual microphone located at $r = 25D$, $\Theta = 40°$, $\Phi = 0°$. Time-frames used for evaluation of OASPL and PSD for the present LES simulation are indicated

The overall sound pressure level (OASPL) of the present jet flows is given in Fig. 7(a). Parts of the burst phenomenon are still included in the LES A time-frame, which results in higher values of the sound pressure level. The peak radiation is at the sideline to slightly forward direction, while at very low angles cancellation effects can be observed resulting in a significant drop of the acoustic power.

The power spectral density (PSD) estimates for the forward, sideline and backward direction are given in Figs. 7(b)–(d). The spectra for the forward angle in Fig. 7(b) almost collapse over a broadband frequency range. Sound emitted at these

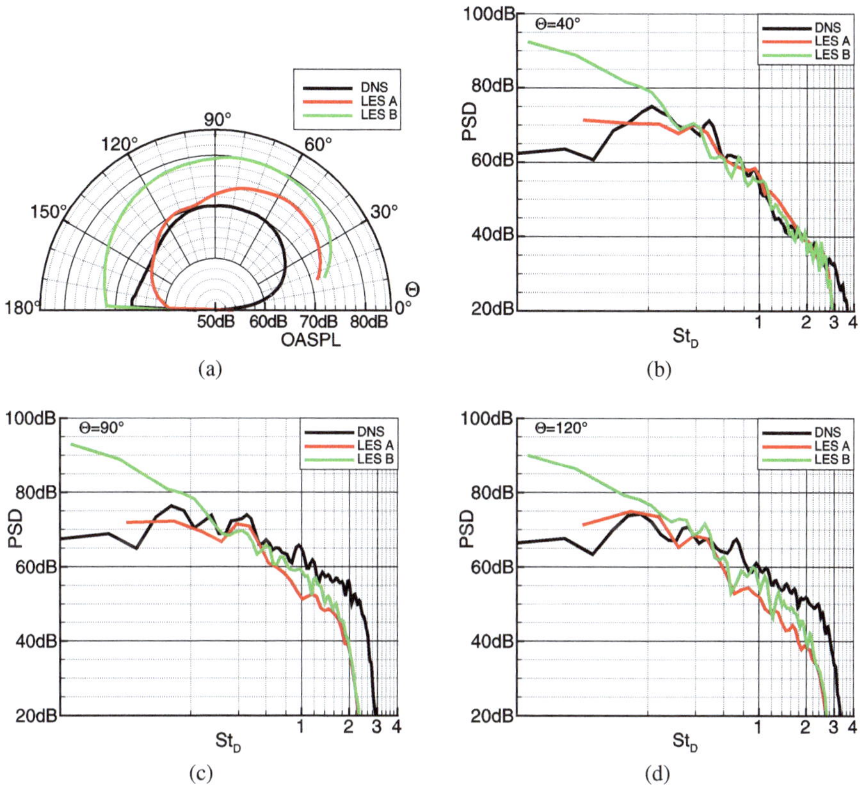

Fig. 7 (a) Overall sound pressure level (OASPL) and (b)–(d) power spectral density (PSD) estimate at azimuthal angles $\Theta = 40°, 90°, 120°$ for the present DNS and LES cases. A and B denote different time-frames for the LES according to Fig. 6

angles is dominated by the large structures in the jet's shear layers, that are well reproduced by carefully performed LES. The LES B case containing the burst phenomenon has large low-frequency contents that do not reproduce the mean statistical acoustic behavior of the flow. However, when the shortened time-frame is used for evaluation (LES A) the emitted acoustical power drops back at low-frequencies, while the decay at higher frequencies is hardly affected. The peak is somewhat underpredicted when compared to that of the DNS, but the amplitude is probably not well captured by the short time evaluation. Furthermore the peak is shifted to somewhat higher frequencies due to the thicker shear layers prescribed for the LES, since thinner layers would not be resolvable. Also the transition to turbulence is different from that of the DNS.

In the sideline and backward direction in Figs. 7(c) and (d), respectively, the jet radiates at lower Strouhal numbers when compared to the forward direction. The lower frequencies for the LES A case are similar to those of the DNS and alike peak locations can be identified. However, the LES fails to predict the sound pressure levels by approximately 10 dB for higher Strouhal numbers, but the slope of the spectra is still comparable to DNS. This emphasizes the inability of LES to capture small structures that dominate acoustic emission in those directions. Maybe more refined subgrid scale (SGS) models could help to improve the situation.

4 Conclusion

Direct numerical and large-eddy simulations of a heavy heated low Mach-number round jet have been successfully conducted. In a second step the sound generation of the jet has been predicted using a hybrid approach solving the acoustic perturbation equations. With this approach aeroacoustic predictions of jet noise are possible in a practicable time frame. Although the computation of acoustic sources from direct numerical simulations is still restricted to low Reynolds number limits. Nevertheless DNS are indispensable for the improvement and validation of large-eddy simulations. The LES presented in this paper was capable to capture the main characteristics of the heated jet and the sound field it emits, but research still needs to focus on improvement of subgrid-scale models with respect to aeroacoustics to enhance results for the sideline and backward direction, where the emitted sound power is underpredicted for high Strouhal numbers.

Massively-parallel supercomputers turned out to be very useful for aerodynamical and aeroacoustical research. The used codes had a adequate performance, but performance tuning is still a difficult task. Mechanisms to map the topology of general computational domains to the topology of the supercomputer are required. Parallel I/O with scalable performance is one of the major challenges for the future, but improvements are necessary since large amounts of data have to be processed in such an aeroacoustic application.

Acknowledgements The authors thank Jülich Supercomputing Centre for the provided computational resources.

References

1. Amielh, M., Djeridane, T., Anselmet, F., Fulachier, L.: Velocity near-field of variable density turbulent jets. Int. J. Heat Mass Transfer **39**(10), 2149–2164 (1996)
2. Bodony, D., Lele, S.: On using large-eddy simulation for the prediction of noise from cold and heated turbulent jets. Physics of Fluids **17**(8), 85–103 (2005)
3. Bodony, D., Lele, S.: Review of the current status of jet noise predictions using large-eddy simulation. AIAA **AIAA-2006-0468** (2006)
4. Bogey, C., Bailly, C.: Direct computation of the sound radiated by a high reynolds number, subsonic round jet. In: CEAS Workshop from CFD to CAA (2002)
5. Bogey, C., Bailly, C.: Noise investigation of a high subsonic, moderate Reynolds number jet, using a compressible large eddy simulation. Theor. Comput. Fluid Dyn. **16**(4), 273–297 (2003)

6. Bogey, C., Bailly, C.: Large eddy simulation of round free jets using explicit filtering with/without dynamic Smagorinsky model. Int. J. Heat and Fluid Flow **27**(4), 603–610 (2006)
7. Bui, T.P., Ihme, M., Schröder, W., Meinke, M.: Analysis of different sound source formulations to simulate combustion generated noise using a hybrid les/ape-rf method. International Journal of Aeroacoustics **8**, 95–124 (2009)
8. Bui, T.P., Schröder, W., Meinke, M.: Numerical analysis of the acoustic field of reacting flows via acoustic perturbation equations. Computers & Fluids **37**, 1157–1169 (2008)
9. Chassaing, P., Antonia, R., Anselmet, F., Joly, L., Sarkar, S.: Variable Density Fluid Turbulence. Kluwer Academic Publishers (2002)
10. Chen, C., Rodi, W.: Vertical Turbulent Buoyant Jets: a Review of Experimental Data. Pergamon Press, Great Britain (1980)
11. Chow, F., Moin, P.: A further study of numerical errors in large- eddy simulations. J. Comput. Phys. **18**, 366–380 (2003)
12. Djeridane, T., Amielh, M., Anselmet, F., Fulachier, L.: Velocity turbulence properties in the near-field region of axisymmetric variable density jets. Phys. Fluids **8**(6), 1614–1630 (1996)
13. Ewert, R., Schröder, W.: Acoustic perturbation equations based on flow decomposition via source filtering. Journal of Computational Physics **188**, 365–398 (2003)
14. Ewert, R., Schröder, W.: On the simulation of trailing edge noise with a hybrid les/ape method. Journal of Sound and Vibration **270**, 509–524 (2004)
15. Fiorina, B., Lele, S.K.: An artificial nonlinear diffusivity method for supersonic reacting flows with shocks. J. Comp. Phys. **222**(1), 246 (2007)
16. Foysi, H., Mellado, J.P., Sarkar, S.: Simulation and comparison of variable density round and plane jets. Int. J. Heat and Fluid Flow **31**, 307–314 (2010)
17. Freund, J.: Compressibility effects in a turbulent annular mixing layer. Dissertation (1997)
18. Freund, J.: Noise sources in a low-reynolds-number turbulent jet at mach 0.9. J. Fluid Mech. **438**, 277–305 (2001)
19. Gröschel, E.R., Schröder, W., Renze, P., Meinke, M., Comte, P.: Noise prediction for a turbulent jet using different hybrid methods. Computers & Fluids **37**(4), 414–426 (2008)
20. Hallberg, M., Strykowski, P.: On the universality of global modes in low-density axisymmetric jets. J. Fluid Mech. **569**, 493 (2006)
21. Hu, F., Hussaini, M., Manthey, J.: Low-dissipation and lowdispersion Runge-Kutta schemes for computational acoustics. J. Comput. Phys. **124**, 177–197 (1996)
22. Ihme, M., Bodony, D., Pitsch, H.: Prediction of combustion-generated noise in non-premixed turbulent jet flames using large-eddy simulation. AIAA Aeroacoustics Conference **2000-2014**, 1–16 (2006)
23. Johansson, S.: High order finite difference operators with the summation by parts property based on DRP schemes. Tech. Rep. 2004-036, it (2004)
24. Kleinman, R.R., Freund, J.B.: The sound from mixing layers simulated with different ranges of turbulence scales. Physics of Fluids **20**(10), 101503 (2008). DOI 10.1063/1.3005823. URL http://link.aip.org/link/?PHF/20/101503/1
25. Koh, S.R., Schröder, W., Meinke, M.: Turbulence and heat excited noise sources in single and coaxial jets. Journal of Sound and Vibration **329**, 786–803 (2010)
26. König, D., Koh, S.R., Meinke, M., Schröder, W.: Two-step simulation of slat noise. Computers and Fluids **39**(3), 512–524 (2010)
27. Kyle, D., Sreenivasan, K.: The instability and breakdown of a round jet variable-density jet. J. Fluid Mech. **249**, 619–664 (1993)
28. Lighthill, M.: On sound generated aeroacoustically: Ii. turbulence as a source of sound. Proc. R. Soc. London Ser. A **222**, 1–32 (1954)
29. Lodato, G., Domingo, P., Vervisch, L.: Three-dimensional boundary conditions for direct and large-eddy simulation of compressible viscous flows. Journal of Computational Physics **227**(10), 5105–5143 (2008)
30. Martin, M., Piomelli, U., Candler, G.: Subgrid-scale models for compressible large-eddy simulations. Theoret. Comp. Fluid Dyn. **13**, 361–376 (2000)
31. Message Passing Interface Forum: MPI: A Message-Passing Interface Standard, Version 2.2. High Performance Computing Center Stuttgart (HLRS) (2009)

32. Mohseni, K., Colonius, T.: Numerical treatment of polar coordinate singularities. J. Comp. Phys. **157**, 787–795 (2000)
33. Popescu, M., Shyy, W., Garbey, M.: Finite volume treatment of dispersion-relation-preserving and optimized prefactored compact schemes for wave propagation. Journal of Computational Physics **210**(2), 705–729 (2005). DOI 10.1016/j.jcp.2005.05.011
34. Ricou, F., Spalding, D.: Measurements of entrainment by axisymmetrical turbulent jets. J. Fluid Mech. **11**, 21 (1961)
35. Ruffin, E., Schiestel, R., Anselmet, F., Amielh, M., Fulachier, L.: Investigation of characteristic scales in variable density turbulent jets using a second-order model. Physics of Fluids **6**(8), 2785–2799 (1994). DOI 10.1063/1.868167. URL http://link.aip.org/link/?PHF/6/2785/1
36. Stanley, S., Sarkar, S.: Influence of Nozzle Conditions and Discrete Forcing on Turbulent Planar Jets. AIAA **38**, 1615–1623 (2000)
37. Tam, C.K.W., Webb, J.C.: Dispersion-relation-preserving finite difference schemes for computational acoustics. J. Comput. Phys. **107**, 262–281 (1993)
38. Tam, C.K.W., Webb, J.C., Dong, Z.: A study of the short wave components in computational acoustics. Journal of Computational Acoustics **1**, 1–30 (1993)
39. The HDF5 Group: Hdf5 reference manual release 1.8.3. Tech. rep. (2009). URL http://www.hdfgroup.org/HDF5/
40. Viswanathan, K.: Aeroacoustics of hot jets. J. Fluid Mech. **516**, 39–82 (2004)
41. Wang, P., Fröhlich, J., Michelassi, V., Rodi, W.: Velocity near-field of variable density turbulent jets. Int. J. Heat Fluid Flow **29**, 654–664 (2008)
42. Zhao, W., Frankel, S., Mongeau, L.: Large Eddy Simulations of Sound Radiation from Subsonic Turbulent Jets. AIAA Journal **39**(8), 161–170 (2001)

Numerical Simulation of Sibilant [s] Using the Real Geometry of a Human Vocal Tract

Kazunori Nozaki

Abstract Speech is the one of the oral function due to fluid dynamics. The sibilant [s], one of dental fricative voice, is assumed to be generated from turbulence around the frontal teeth. Large Eddy Simulation with taking the real morphological vocal tract into account was demonstrated. The mean flow of the sibilant [s] separates from the tip of the upper anterior teeth and induces half plane jet along the lower anterior teeth. ΔP on surface of the anterior teeth is assumed to be somehow linked to the location of the high value Powell sound source.

1 Introduction

In speech science, there are interdisciplinary approaches to understand the physical mechanisms of sound production. There are two significant acoustic properties of the human voice, resonance and sound source. Human voices are articulated in the airway that has unique characteristics in the resonance given from the human vocal tract that consists of vocal cords, a larynx, pharynges, a nasal cavity, an oral cavity, teeth and lips (Fig. 1). The articulation physically perform the change of the geometry configuration of the airway in accordance with phonemes in syllables. The specific resonance and harmonics are generated by the specific geometry configuration in the human vocal tract. The vocal cords, an oral cavity, teeth and lips can generate sound. The place or region where generates sound is called "sound source". The sound source gives the energy to the human vocal tract. Fant (1970) proposed the source-filter theory that enables to deal with resonance and sound source separately [1]. The acoustical analyses has been applied to examine the speech disorders. The signal processing has been mostly applied to detect the each phoneme and also

Kazunori Nozaki
The Center for Advanced Medical Engineering and Informatics, Osaka University, 5-1 Mihogaoka Ibaraki, Osaka, Japan
e-mail: nozaki@dentgrid.org

M. Resch et al. (eds.), *High Performance Computing on Vector Systems 2010*,
DOI 10.1007/978-3-642-11851-7_10,

quantify the change of the voice. This article focus on the turbulence sound which is the one of the consonant articulated in the oral cavity.

1.1 Signal Processing for Consonants

The quantification of the turbulence sound has been performed several purposes [2–6]. An ensemble averaging discrete Fourier transform, spectral moments or fractal dimensions analyses are thereby mostly utilized [2–5]. Those methods can consider the unsteady signal which assumed to be generated from turbulence [6]. The characteristics of sound source, however, are not clear, for example, source locations, strength and the characteristics frequency.

1.2 Sound Induced Flow

There is the flow induces sound, for example, jet, explosion, duct, turbine, compressor, propeller sound, aeolian, edge, pipe, cavity tone, sonic boom, etc. The basic model and mechanism of the sound production has been studied since Lighthill (1952) propose Lighthill stress tensor with the wave equation in the right hand that is elicited by the compressible Navier Stokes equation without any abbreviation [7–11]. The mathematical model seems to enable scientists and dentists to predict the speech sound after some alterations of geometries in vocal tracts. The mathematical model still requires some approximations to apply to solve sound generated by the flow in complicated geometries.

1.3 Sibiant [s] in Dental Treatments

There are growing expectations of researching and developing fundamental treatment methods of speech disorders, which are directly linked to Quality Of Life (QOL). In particular, the dental fricative called sibilant [s], has different characteristics of its sound production mechanism with the one of vowels. The sound source of the sibilant [s] is downstream obstacle in the oral cavity, where turbulence is thought to be dominant, although the vowel's sound source is the vibration of vocal cords by airflow pressure [12, 13]. That is to say, frontal teeth are thought to be the obstacles against the laminar or turbulence flow so far. Modification of the features of oral cavity is often occurred on changing spatial positioning of jaws on the purpose of maxillary orthodontic therapies [14], prosthetic treatments [15] and inserting sports mouth guards [16]. These treatments may affect not only the characteristics of the resonance, but also the location and magnitude of sound sources of the sibilant [s].

1.4 Computational Analyses for Sibilant [s]

The sibilant [s] can be analyzed by considering turbulence and vortex sound, both of which are studied in terms of numerical analyses such as computational fluid dynamics (CFD) and computational aero acoustics (CAA) [17–19]. Both CFD and CAA require a computational mesh that is built by a projection of the morphological shape of the oral cavity. The anterior teeth shape in particular has a prominent effect on the sound generation of the sibilant /s/, and the shapes of these teeth thus need to be taken as precisely as possible to perform CFD and CAA.

1.5 Complicated Morphology of Vocal Tracts

It is necessary to obtain geometries of both soft tissues, such as lips chests, a tongue, etc, and hard tissues, teeth simultaneously. The method of taking those geometries has been taking cross-sectional images of the oral cavity during utterance of the sibilant [s]. As this method is restricted by the total acquisition time during the utterance of it, it is not appropriate choice to use a normal Computed Tomography (CT). The normal CT, moreover, can not take the narrow (1-mm) space between upper and lower teeth is formed during the sibilant [s] is pronounced [20]. So that, three dimensional geometry of the narrow space needs to be measured by some specific way. It is also difficult to generate the computational mesh of the oral cavity, especially, the narrow space. The obstacle sound source, the sibilant [s], is explained that the impact of turbulence on the obstacle's surface generates sound [21–24]. It is natural to think that the distribution of the source would be changed, if the shape of teeth surface was altered, because the sound source is distributed around the surface of teeth. It is, moreover, well known that the sound generated from the interaction between the surface and the vortexes is louder than the one from the interaction among vortexes themselves in the flow field.

1.6 LES and Aeroacoustics

Compressible Navier-Stokes equations using Direct Navier-Stokes (DNS) method certainly can predict broadband noise, however, it is clear that impractical number of mesh and extremely fine mesh are necessary. More practical and promising approach for aeroacoustics is to apply separation methods. The separation method has been used because of the great disparity of levels and length scales. With this method, the near-field flow field is computed to get velocity fluctuations that are used to compute the sound sources for a separate computation of the far-field pressure fluctuations [25]. CFD methods have faced many difficulties to deal with the vortex unready change. As acoustical problems are highly time dependent, Reynolds-averaged (time-averaged) Navier-Stokes (RANS) equations are not adequate for the time series analysis. Large Eddy Simulation (LES) method is, contrary, suitable to deal with these problems. LES method uses Sub-Grid Scale (SGS) model to consider vortexes affects for smaller grids than originals. As the LES uses time-independent and space averaged model, it is suitable to analyze time depending vortexes. However, the accuracy of the LES depends on the type of mesh, such as hexahedron, tetrahedrite, prism, etc. Especially, for the zones in turbulence boundary layer, the mesh size in a normal direction on the boundary surface needs to be as small as possible. The quality of boundary mesh affects the result of the LES, because the sibilant [s] is thought to be generated from frication on the wall. Lighthill (1952) showed the far field sound propagation could be computed. His theory was derived by transformation of compressible Navier-Stokes equation to the wave equation in the left term and sound sources term in the right [7]. The aim of this study is to explore the sound production mechanisms of the sibilant [s] in case of the real geometry of a human oral cavity by performing CFD and CAA.

2 Materials and Methods

In order to acquire the three dimensional geometry of the narrow space around the anterior teeth within the duration of utterance as long as possible, Cone Beam type of CT scanner is adequate. The oral cavity shape of the sibilant [s], therefore, was

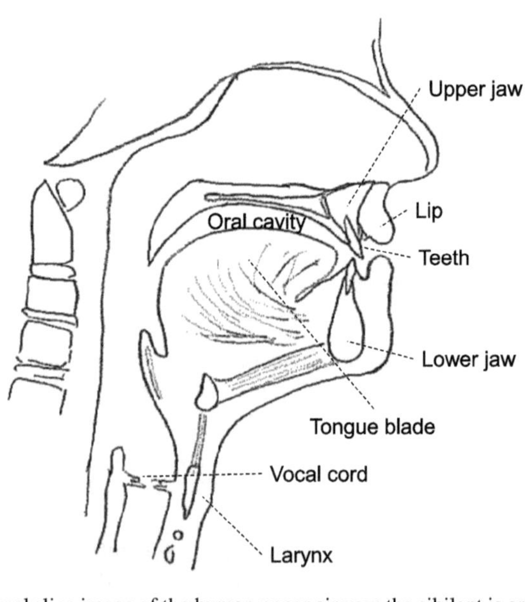

Fig. 1 The orthogonal slice image of the human upper airway: the sibilant is articulated the region of the oral cavity including teeth and lips

obtained by using Cone Beam CT (CBCT) scanner (Hitach medico Corp.) that can take 512 slices by 512×512 pixels within 18 seconds. Those slices' images form a volume by the volume rendering. The size of each volume is 0.1 mm although the normal medical use CT only take the volume by 0.3 mm. The oral cavity only could be extracted with two threshold CT values. Marching Cubes method could construct surfaces of the oral cavity by using RealIntage (KGT Corp.). The surface was converted from Stereolithography Interface Format (STL) format to Non-Uniform Rational B-Spline (NURBS) format. Computational mesh was constructed with a multi block of hexahedral elements by using Gridgen (Pointwise Corp.). The mesh was 2.64 million mesh (2.64 M) at first. It is important to examine the mesh size effects that is required to determine the mesh size is enough small to calculate vortexes. Each element, so that, was divided into n elements equally by using our custom program, where n means integer, such as $1, 2, 3 \ldots, n$.

It was considered that LES consumes a large amount of computational resources. LES is the one of the modeling method for turbulence. FrontFlow Blue (C. Kato, et al.) was chosen to perform LES because the source code is freely opened and not only prepared for parallel computation by Message Passing Interface (MPI), but also customized for vector super computers. FFB was performed in SX8-R (NEC Corp.). FFB has both 2nd order accuracy of time and scale. In our study, the dynamic smagorinsky model (DSM) was used as Sub-Grid Scale (SGS), because DSM can calculate the SGS stress even on the wall by using the smagorinsky model parameter obtained by the average velocity of Grid Scale (GS). An explicit method was performed for time-marching. The time delta was 1.0×10^{-6}. In Flow condition was set 6.5 m/s at Ymax and Out Flow conditions were set at Ymin, Zmax, Zmin, Xmax and Xmin (Fig. 2). The Reynolds number was set to 6708 this time.

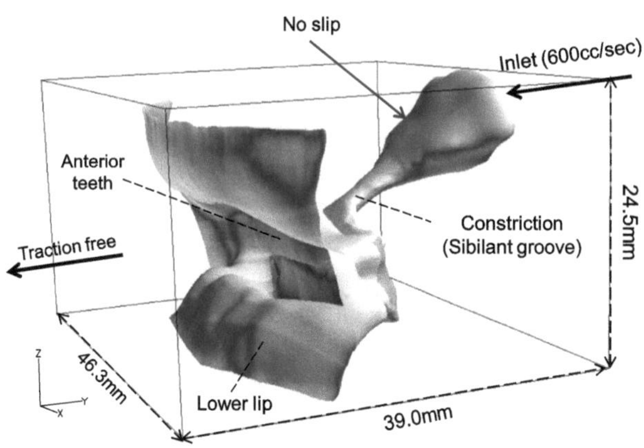

Fig. 2 Boundary condition of the sibilant [s] LES simulation: Uniform velocity of 6.4 m/s (Inlet), Traction free (Outlet) and No slip (Wall)

3 Result

The mesh modeling procedure was shown in Fig. 3. It was found to be possible to get the whole of the vocal tract within the persistent utterance time. It was not possible to acquire the volume of the vocal tract at the single CT value (Hounsfield value) of air because CBCT's strength of X-ray is weaker than the conventional one, though better for the risk of X-ray exposures. Two different CT values were decided to be considered as the air. NURBS surface was created by using the STL data obtained by the vocal tract volume data. Surface meshes were firstly created and then the volume mesh was constructed. This mesh consisted of 2.64 million elements. The elements of 2.64 million mesh (2.64 M) were divided to be 8 times and 27 times larger mesh size.

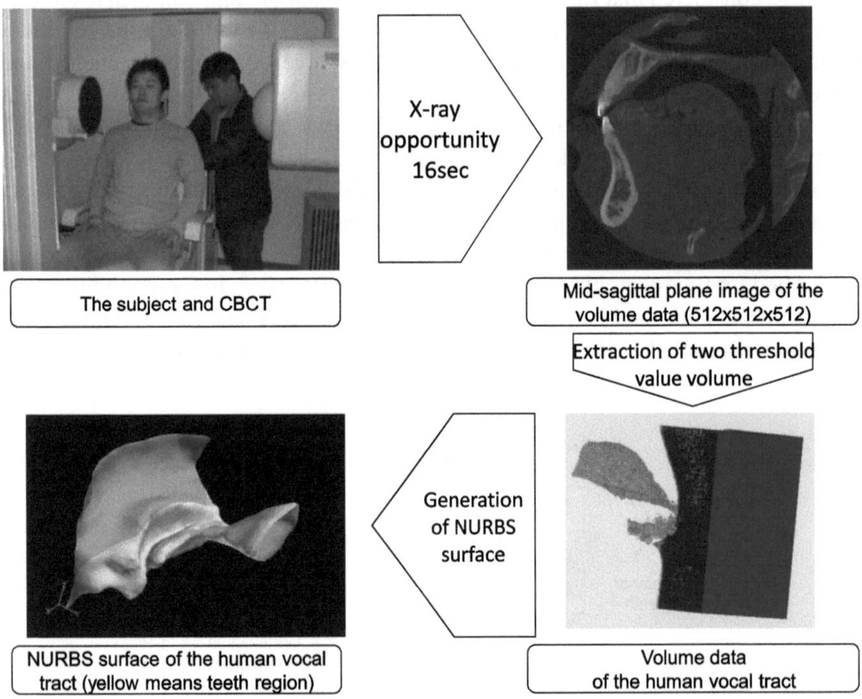

Fig. 3 The procedure to obtain the NURBS surface to construct the hexahedral mesh: Image picture of CBCT, the mid-sagittal image, the segmented volume and the NURBS surface of the subject's vocal tract

The result of LES analyses by FrontFlow/Blue was shown in Fig. 4. The averaged velocity magnitude of 4,000 time steps in 4 ms in both cases of the coarse and fine meshes in Fig. 5 (up). In both cases, the mean flow separates from the tip of the upper anterior teeth and induces half plane jet along the lower anterior teeth. The root mean square (rms) values in the case the fine mesh has different characteristics

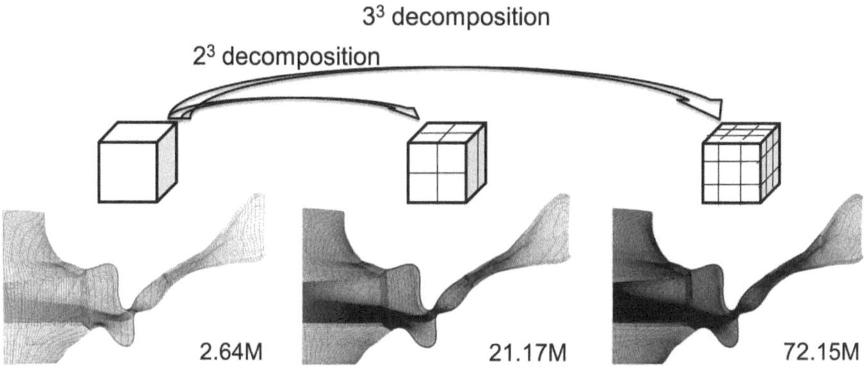

Fig. 4 Three levels of the human vocal tract's mesh: Coarse mesh (2.64 M), Fine mesh (21.17 M), Very fine mesh (72.15 M)

on the jet from the coarse mesh. The level of the rms values in the fine mesh are much higher than the coarse mesh. As the rms values is linked to the turbulence energy in LES analyses, the fine mesh could calculate the ρuv and v_{SGS}, where ρ is air density, uv is Reynolds stress and v_{SGS} is dynamic coefficient of viscosity with SGS filtered. The very fine mesh (71.15 M) result in Fig. 6 backed up the accuracy if the LES SGS filtered result.

The sound source that is called Powell sound source, $\rho\nabla\cdot(\omega\times u)$. was visualized in Fig. 7. The contour on the anterior teeth indicates $\triangle P$ which means the magnitude of pressure change on the surface. Lighthill (1963) and Howe (1998) mentioned that $\triangle P$ should not be zero on the surface in vorticity production. The anterior teeth regarded as sound source requires nonzero $\triangle P$ area on its surface. The iso-surface shows the location of high value Powell sound sources. The contour of the iso-surfaces represents the magnitude of vorticity. It was cleared that diffusion of vortexes is relatively stronger than vorticity, because the value of vorticity is not significantly high on the surface of the Powell sound sources. $\triangle P$ on surface of the anterior teeth is assumed to be somehow linked to the location of the high value Powell sound source.

4 Discussion

The complicated vocal tract has been targeted for the numerical simulation of the sibilant [s] in this article. The reliability and the other approaches should be discussed to have any conclusion. Sound propagation, reflection and resonance in the cavity has not considered in this article. The integrated simulation is required to take acoustics into account. The key point to achieve the integrated simulation is hereby the total amount of storage size and also high speed network connecting the storage farm to computational farm.

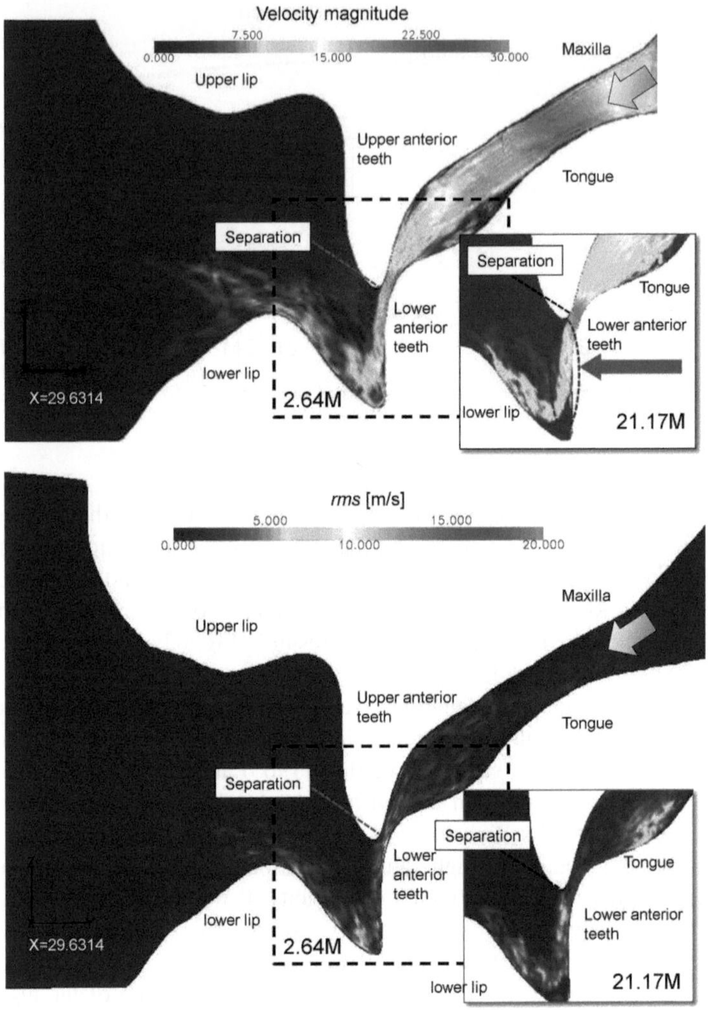

Fig. 5 The result of the averaged (up) and the root mean square (low) values of flow velocity magnitude of the human vocal tract in case of the coarse mesh (2.64 M) and the fine mesh (21.17 M)

4.1 Validity of Real Morphological Geometry

The numerical simulation with taking the real morphological vocal tract into account was demonstrated in this article. This should be considered as one of possible approaches to study the sound production of the sibilant [s]. This numerical simulation of the sibilant [s], however, produce the useful indication supported by the real geometry taken from the human vocal tract. The results such as the averaged flow, rms and the location of sound sources, at least, showed the difference phenomena from the result by using some simplified model of the sibilant [s] [19]. The sim-

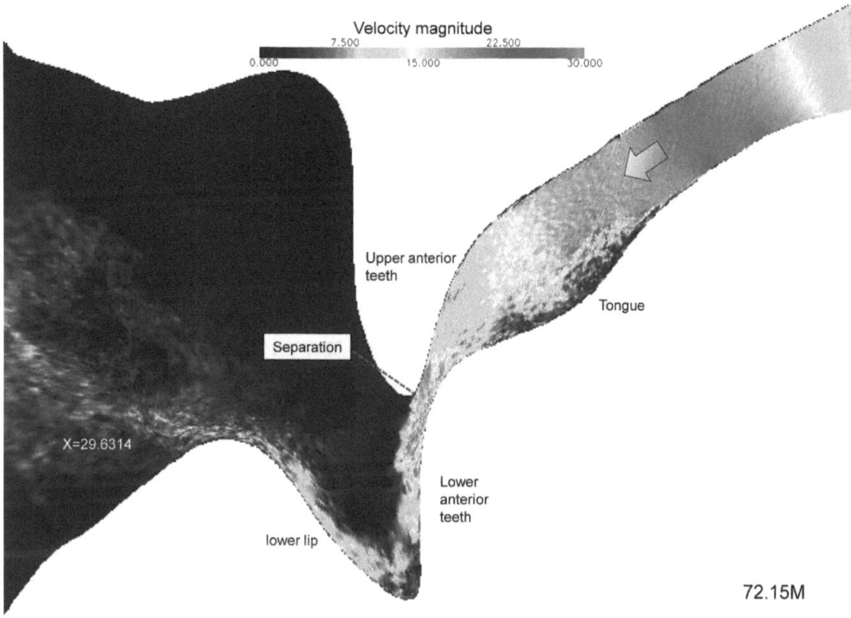

Fig. 6 The result of the averaged flow velocity magnitude of the human vocal tract in the very fine mesh (71.15 M)

plified model has many positive reason to be taken as a first approach to study the sibilant [s]. It is, for example, easy to understand the relationships and interactions between flow and geometries, moreover, to compare the previous reports that sometimes suggest many important criteria. It is nevertheless necessary to explain how or why the simplified geometry represent the human vocal tract. The only solution of this question is to perform both of geometries complementary in the future.

On the other hand, the accuracy of representing the real morphological features in this article is not perfect currently, because it requires tremendously efforts to built the multi-block hexahedral mesh of the complicated vocal tract including teeth. The tolerance of the hexahedral mesh is not best to apply the complicated geometry normally, however, the ability of keeping the accuracy near walls is one of the best. The hexahedral mesh was therefore chosen in this article.

4.2 Aeroacoustic Analyses

The theory of aeroacoustics has been established since 1960s–1970s by Lighthill (1952), Powell (1964), Howe (1998), etc. Howe, et al. (2005) introduce the mathematical solution of the acoustics of the sibilant [s] for the first time [26]. They explained the sound production mechanism by using compact green function in

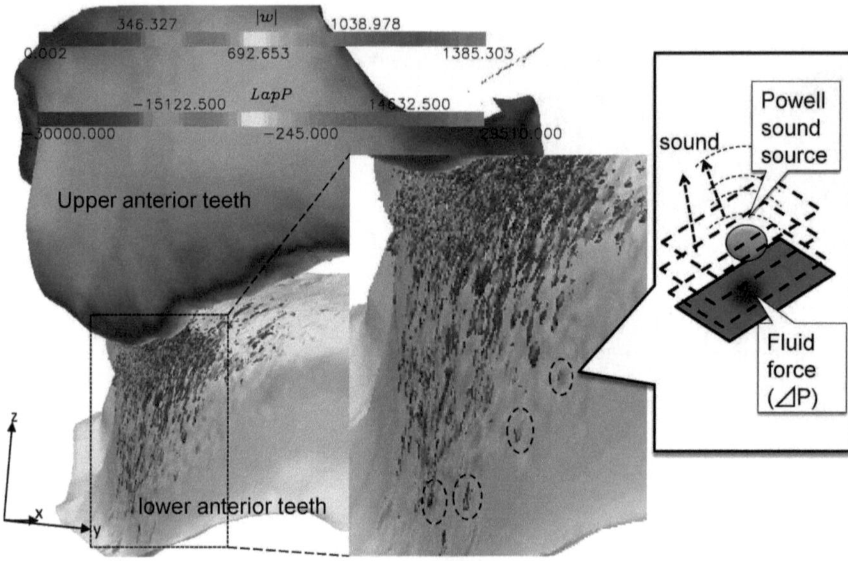

Fig. 7 Powell sound source in the fine mesh (21.17 M): The Fluid force on the surface may generate sound waves subsequently to Powell sound source wave

two dimensional simplified constriction between teeth. Shadle (1985, 1990) showed the experimental setup to produce the sibilant [s] sound which has been utilized as "benchmark" setup [22, 24]. Their significant important studies should support the dental treatments to predict or answer the problem of the production of the sibilant [s]. The simplified geometry is, however, not best choice to answer the request because the shape of the teeth or the vocal cavity is not well linked to the real morphology. Thus, the complicated setup, such as this article's, is needed, at least for the prostheses, etc., although the high performance computing (HPC) is necessary. In order to produce the predictive sibilant [s] sound, we need to perform the separated solution of aeroacoustical simulation that has two phase, one is CFD (LES) and next is CAA. The lighthill tensor is required by LES in time course, and then used in Helmholtz equation solver. The essential boundary condition should be impedance wall and non-reflective condition.

4.3 Requirements for High Performance Computing

The computation scheme of LES request the flat topology among CPUs to the computation environments. The incompressible Navier Stokes requires to solve Poisson equation to obtain the value of pressure in LES. Most of computation time of LES is consumed by Poisson solver. Poisson equation is normally solved numerically by solving simultaneous equations. The machines that is faster to solve the simultaneous equations is therefor used.

The another aspect of HPC is how to accelerate the integrated simulation such as CFD and CAA. The total performance of the computational environment is significant in this case because the file transfer time, usability of large scale storages and computational power balance effect the total computation time.

5 Conclusion

The numerical simulation with taking the real morphological vocal tract into account was demonstrated. The mean flow of the sibilant [s] separates from the tip of the upper anterior teeth and induces half plane jet along the lower anterior teeth. $\triangle P$ on surface of the anterior teeth is assumed to be somehow linked to the location of the high value Powell sound source.

Acknowledgements I would like to cordially appreciate giving chance to take CBCT to Dr. Tamagawa, Dr. Miura and DEMIC dental modern. Prof. Kato, Dr. Guo Dr. Mizushima Mr. Yamade gave tremendously support and advices to me for this research. I appreciate cordially managing HPC environments to Prof. Shimojo, Dr. Higashida. Software development of the mesh scaling owed to Mr. Maeda. This research was supported by MEXT(40379110).

References

1. Fant, G.: Acoustic Theory of Speech Production With Calculations based on X-Ray Studies of Russian Articulations. Description and Analysis of Contemporary Standard Russian, pp. 15–26, Mouton, The Hague, Paris (1970)
2. Oppenheim, V.A. and Schafer, W.R.: Digital Signal Processing. Prentice-Hall, Englewood Cliffs, New Jersey (1975)
3. Forrest, K., Weismer, G., Milenkovic, P., and Dougall, R.N.: Statistical analysis of word-initial voiceless obstruents: preliminary data. J. Acoust. Soc. Am., **84**, 115–123 (1988)
4. Harrington, J., Cassidy, S.: Techniques in speech acoustics. Kluwer Academic Publishers, Netherland (1999)
5. Jongman, A., Wayland, R. and Wong, S.: Acoustic characteristics of English fricatives. J. Acoust. Soc. Am., **108**, 1252–1263 (2000)
6. Pitsikalis, V. and Maragosa, P.: Analysis and classification of speech signals by generalized fractal dimension features. Speech Communication, **51,12**, 1206–1223 (2009)
7. Lighthill, M.J.: On sound generated aerodynamically I, General theory. Proc. R. Soc. London, **A 1211**, 564–587 (1952)
8. Lighthill, M.J.: II. Introduction Boundary Layer Theory. In: Rosenhead, L. (ed.) Laminar Boundary Layers, pp. 46–109. Oxford University Press, Oxford (1963)
9. Powell, A.: Theory of vortex sound. J. Acoust Soc. Am., **33**, 177–195 (1964)
10. Möhring, W.: On vortex sound at low Mach number. J. Fluid Mech., **85**, 685–691 (1978)
11. Howe, M.S.: Acoustics of Fluid-Structure Interactions. Cambridge University Press, Cambridge (1998)

12. Pelorson, X., Hirschberg, A., Wijnands A.P.J. and Bailliet, H.: Description of the Flow through In-vitro Models of the Glottisduring Phonation. Acta Acoustica, **3**, 191–202 (1995)
13. Zhao, W., Zhang, C., Frankel, S.H. and Mongeau, L.: Computational Aeroacousics of Phonation, Part I: Computational Method and Sound Generation Mechanisms. J. Acoust Soc. Am., **112**, 2134–2146 (2002)
14. Hassan, T., Naini, F.B. and Gill, D.S.: The effects of orthognathic surgery on speech: a review. J. Oral Maxillofac. Surg. **65**, 2536–2543 (2007)
15. Hamlet, S., Geoffrey, V.D. and Bartlett, D.M.: Effect of a dental prosthesis on speaker—Specific characteristics of voice. J. Speech Hear. Res., **19**, 639–650 (1976)
16. Banky, J.: Mouthguards, dental injury and problems: On-field management. Journal of Science and Medicine in Sport, **3, 2**, 5–11 (2000)
17. Nozaki, K., Akiyama, T., Tamagawa, H., Kato, S., Mizuno-Matumoto, Y., Nakagawa, M., Maeda, Y. and Shimojo, S.: The first grid for oral and maxillofacial region and its application for speech analysis. Methods Inf. Med. **44**, 253–256 (2005)
18. Nozaki, K., Tamagawa, H. and Shimojo, S.: Prediction of dental fricative sound by Lighthill-Curle equation. In: Takada, K. (ed.) In silicon Dentistry, pp. 155–158. MEDIGIT, Osaka (2008)
19. Van Hirtum, A., Grandchamp, X., Pelorson, X., Nozaki, K. and Shimojo, S.: LES simulation and 'in-vitro' experimental validation of flow around a teeth-shaped obstacle. Int J. Appl. Mech. (in press)
20. Suda, T., Nozaki, K., Higashida, M. and Baba, K.: The Complementary Registration Method for Vocal Tract of Sibilant /s/. In Proceedings of SAINT 2010. (in press)
21. Stevens, K.N.: Airflow and turbulence noise for fricative and stop consonants: Static considerations. J. Acoust. Soc. Am., **50**, 1180–1192 (1971)
22. Shadle, C.H.: The acoustics of fricative consonants. MIT Tech. Rep. Cambridge, MA (1985)
23. Stevens, K.N.: Acoustic Phonetics, pp. 379–412. MIT Press, Cambridge MA (1988)
24. Shadle, C.H.: Articulatory-acoustic relationships in fricative consonants, In: Hardcastle, W.J. and Marcha, A. (eds.) Speech Production and Speech Modeling, pp. 187–209. Kluwer Academic Publishers, The Netherlands (1990)
25. Obrai, A.A. and Pinsky, P.M.: A residual-based finite element method for the Helmholtz equation. Int. J. Numer. Meth. Engng, **49**, 399–419 (2000)
26. Howe, M.S. and McGowan, R.S.: Aeroacoustics of [s]. Proc. R. Soc. A, **461**, 1005–1028 (2005)

Identification of Anisotropic Elastic Material Properties by Direct Mechanical Simulations: Estimation of Process Chain Resource Requirements

Ralf Schneider

Abstract In this work the effort necessary, to derive a linear elastic, inhomogeneous, anisotropic material model for cancellous bone from micro computer tomographic data via direct mechanical simulations, is analyzed. First a short introduction to the background of biomechanical simulations of bone-implant-systems is given along with some theoretical background about the direct mechanics approach. The implementations of the single parts of the simulation process chain are presented and analyzed with respect to their resource requirements in terms of CPU time and amount of I/O-data per core and second. As the result of this work the data from the analysis of the single implementations are collected and from test cases, the data were derived from, they are extrapolated to an application of the technique to a complete human femur.

1 Introduction

The global aim of the biomechanical research done at HLRS is the support of diagnostic decisions with respect to type and positioning of implants, used for certain pathologies. In addition the healing process as well as the possibility of mechanical complications should be predicted. The pathologies currently analyzed at HLRS are abdominal aortic aneurysms and the postoperative behavior of intramedullary implants as they are used for the fixation of fractured bones.

For static Finite-Element simulations of bone implant-systems three prerequisites are essential. (1) The bone geometry, (2) the forces acting on the bone structures which includes body—as well as muscle forces and (3) a detailed modelling of the inhomogeneous, anisotropic material distribution of the bony structures.

Ralf Schneider
High Performance Computing Center Stuttgart (HLRS), Nobelstraße 19 70569 Stuttgart, Germany
e-mail: schneider@hlrs.de

M. Resch et al. (eds.), *High Performance Computing on Vector Systems 2010*,
DOI 10.1007/978-3-642-11851-7_11,

Since clinical imaging data, as recorded by Computer Tomography (CT) or Magnet Resonance Imaging (MRI), are the only data which are available from living bones all three prerequisites have o be fulfilled by the detailed analysis of these data.

To separate the micro structures of cancellous bone, which have a diameter of ≈ 0.1 mm an isotropic resolution of ≈ 0.02 mm would be necessary. Since this resolution is approximately 20 times higher than the one which can be achieved with clinical dual source CT scanners an additional modelling step has to be introduced to enable mechanical simulations on the resolution level of clinical CTs.

The chosen approach to this additional modelling step is, to scan cancellous bone specimens with a technical micro CT scanner which is able to separate the trabecular micro structures and with a clinical CT scanners. Then simulate virtually cut cubes, so called representative volume elements (RVEs), with the directly modelled micro structures under compression and shear loads and derive the material constants of the RVEs on the resolution level of clinical CTs with the standard mechanics approach (see Sect. 2.1). Afterwards a matching of the clinical CT density field and the field of the calculated material constants can be done with multivariate analysis methods, such as cluster analysis for example, to find a transfer function which enables the calculation of anisotropic, elastic material constants directly from clinical CT data. The development cycle for the chosen approach is shown in Fig. 1.

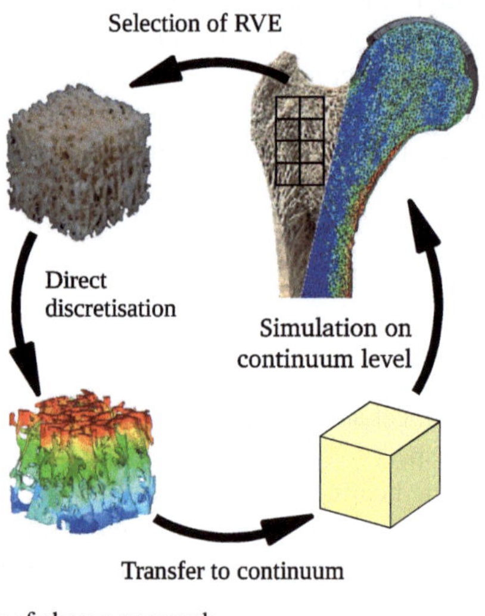

Fig. 1 Development cycle of chosen approach

2 Material and Methods

In this section the subject matter to this study, a process chain for the derivation of an anisotropic, linear-elastic material model for cancellous bone based on clinical CT-data is described. For the better understanding of the theoretical background a short introduction to the direct mechanics approach is given, followed by the description of the single chain links, coupled together in the complete process chain.

2.1 Theoretical Background—Standard Mechanics Approach

With the standard mechanics approach a procedure is described which is often used as the starting point to the application of homogenization theory [1]. Aim of the approach is the determination of averaged elastic properties of micro structured materials on the continuum level. One perquisite which is essential to this approach is, that the material properties of the micro structures in the analyzed RVE are known. In this study the material of the bone micro structures is considered to be homogeneous with isotropic behavior according to [2].

To calculate the averaged macroscopic stress and strain acting over a RVE from the local microscopic strains and stresses in the micro structures the integral averages are used.

$$\overline{\varepsilon}_{ij} = \frac{1}{|V_{RVE}|} \int\limits_{V_{RVE}} \varepsilon_{ij}\, dV_{RVE} \tag{1}$$

$$\overline{\sigma}_{ij} = \frac{1}{|V_{RVE}|} \int\limits_{V_{RVE}} \sigma_{ij}\, dV_{RVE} \tag{2}$$

With: $\overline{\varepsilon}_{ij}$ and $\overline{\sigma}_{ij}$ the mean strain and stress tensors over the RVE, ε_{ij} and σ_{ij} the local strain and stress tensors and V_{RVE} the volume of the RVE.

To connect the averaged macroscopic strain, applied to the RVE, with the resulting averaged macroscopic stress one defines

$$\overline{\sigma}_{ij} = \overline{C}_{ijkl}\overline{\varepsilon}^{kl} \tag{3}$$

with $\overline{C}_{ijkl}$ the so called effective stiffness of the RVE which is a tensor of rank four.

Furthermore, to connect the local microscopic strains with the averaged macroscopic strain, the local structure tensor M_{ijkl}, which is also a tensor of rank four, is defined by

$$\varepsilon_{ij} = M_{ijkl}\overline{\varepsilon}^{kl} \tag{4}$$

The development of the correlation between the elastic properties of the microscopic structures to the averaged elastic properties of the RVE is started from the generalized Hooke's law on the microscopic level

$$\sigma_{ij} = C_{ijkl}\,\varepsilon^{kl} \tag{5}$$

Integrating both sides of Eq. 5 over the RVE yields

$$\frac{1}{|V_{RVE}|} \int\limits_{V_{RVE}} \sigma_{ij}\, dV_{RVE} = \frac{1}{|V_{RVE}|} \int\limits_{V_{RVE}} C_{ijkl}\,\varepsilon^{kl}\, dV_{RVE} \tag{6}$$

Substituting the left hand side with Eq. 2 and ε^{kl} with Eq. 4 we get

$$\overline{\sigma}_{ij} = \frac{1}{|V_{RVE}|} \int_{V_{RVE}} C_{ijmn} M^{mn}{}_{kl} \, dV_{RVE} \, \overline{\varepsilon}^{kl} \tag{7}$$

Comparing Eq. 7 to Eq. 3 one recognizes the relation

$$\overline{C}_{ijkl} = \frac{1}{|V_{RVE}|} \int_{V_{RVE}} C_{ijmn} M^{mn}{}_{kl} \, dV_{RVE} \tag{8}$$

from which the effective stiffness of the RVE $\overline{C}_{ijkl}$ can be calculated, if the local elastic properties and the function of the local structure tensor of the microscopic structure are known.

If the microscopic structure of the bone material is modelled directly, the function of the local structure tensor can be calculated via Eq. 4 from the local strain field in the microscopic structure which can be obtained by Finite-Element simulations.

For the analysis done in this study a human femoral head, removed during the insertion of a total hip endoprosthesis, was scanned with a micro CT at the DLR Stuttgart—Intitut für Bauweisen und Konstruktionsforschung with a isotropic voxel size of 0.018 mm edge length. From this scan accurate models of the bone micro structure, like the one shown in Fig. 2, are obtained.

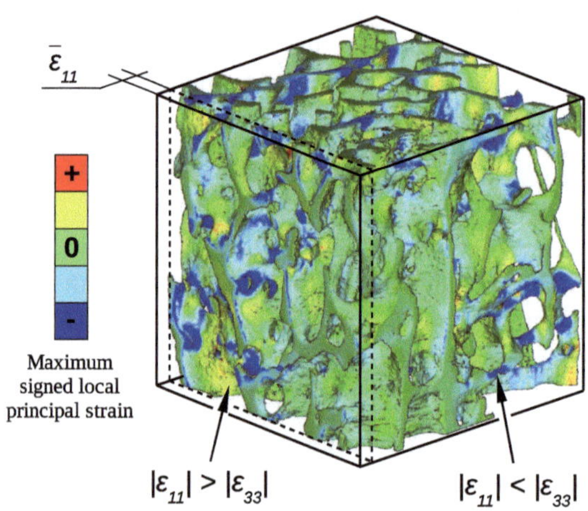

Fig. 2 Tabecular structures of cancellous bone modelled from micro CT

If the technique described above is executed over a complete human bone, a transfer function can be found which connects clinical CT data to the field of the effective stiffness tensors of a certain RVE size.

So (8) becomes

$$\overline{C}_{ijkl} = \overline{\mathbf{C}}\,(\overline{\mathbf{x}}) \tag{9}$$

with $\overline{\mathbf{x}} = [\overline{x}\,\overline{y}\,\overline{z}]$ the macroscopic coordinates which describe the RVE centres at the considered RVE size.

2.2 Process Chain Description

In Fig. 3 an overview of the general approach with all necessary steps is given. The process chain which is analyzed for resource requirements in this study is shown on the left. It starts with the geometry setup and ends with the minimization of the shear-coupling coefficients so it can basically subdivided into four separate chain links which are described in the following.

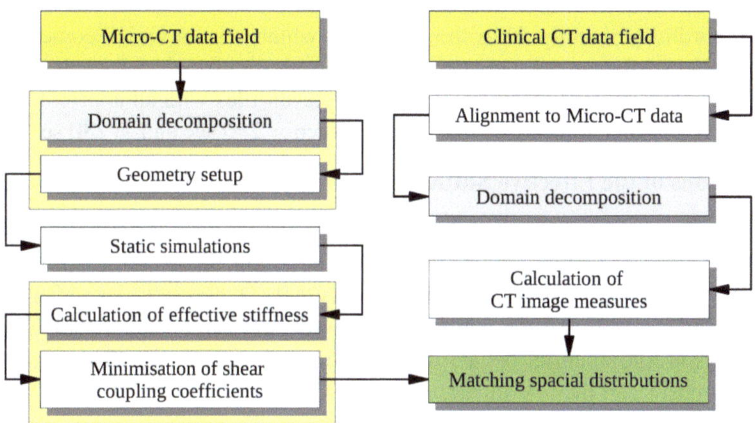

Fig. 3 Process chain description: General approach

Geometry Setup

For the geometry setup two algorithms are implemented. The first one is a voxel based technique which simply transforms voxels from the micro-CT data to FEs according to a greyscale threshold. To avoid unconnected regions in the model after the meshing a check has to be performed which analyzes the meshed regions for connection to the RVE boundaries. The advantages of this algorithm are its robustness and speed. Its disadvantage is that the surfaces of the trabecular micro structures are "bricked".

The second implementation for the geometry setup starts with an advancing fronts algorithm based on [3]. The algorithm is modified in a way that it can directly triangulate isosurfaces in a scalar field such as a micro-CT data field. Its advantage is the generation of a perfectly smooth surface mesh with triangles of high enough quality to generate a tetrahedral volume mesh, suitable for FE simulations, directly from the surface mesh. The disadvantage of the algorithm is an approximately 10 times higher CPU time an a less robustness compared to the voxel mesher.

The decision which algorithm to use for the geometry setup isn't final yet since the analysis whether it is necessary to use the more accurate Advancing Fronts or the fast and resource preserving voxel mesher is not finalized. In Sect. 3 the values for the voxel based mesher are given since its resource measurements are more accurate at the moment.

Static Simulations

To determine the field of the local structure tensor for a RVE six load case simulations have to be done. Since the aim is to determine linear-elastic material data for the solution of the load cases only one linear step is necessary per load case. The average model size is estimated to 300.000 degrees of freedom (DOF).

For the solution of the FE problems, the commercial FE-solver ABAQUS[1] is used. For the model setup it allows to define the boundary conditions and solver parameters independently from the geometry which results in 1 Geometry input file with $\approx$ 18 MB and 6 Load-case description files with $\approx$ 25 kB each. All static simulations for one RVE produce 18 relevant Result files with an approximate size of 720 MB within a total wall clock time between $\approx$ 250 sec and $\approx$ 280 sec.

Calculations of the Effective Stiffness

Due to the usage of the FE-solver ABAQUS, the calculation of the effective stiffness of a RVE consists of two major steps. First the strain fields of the six load case simulations have to be extracted from the binary ABAQUS result files. After that the local structure tensor and the volume of each element can be calculated. Together with the isotropic material behavior of the micro structures the field of the local structure tensor can then be integrated according to Eq. 8 to the effective stiffness of the RVE.

The extraction of the strain fields from the result files is referred to as a separate step because this part of the effective stiffness calculation takes about 270 sec wall clock time and mainly depends on the used FE-solver while the actual calculation of the local structure tensors and the integration of the effective stiffness take only 4 sec of wall clock time. In addition the results of the data transformation introduce 24 new files to the process which cover $\approx$ 360 MB of storage size.

Transformation of the Effective Stiffness

After the last step the effective stiffness tensors of the RVEs are represented in the global coordinate system where they are dense in general. If one wants to determine material symmetries, the representation coordinate system has to be rotated in

[1] ABAQUS is a trademark of SIMULIA, Rising Sun Mills, 166 Valley Street, Providence, RI 02909-2499

order to find the direction where the normal to shear and shear to shear coupling coefficients are of minimum value.

This step is added to the process chain because in the literature cancellous bone is said to be orthotropic [4, 5] but studies done with the presented process chain on a primed cancellous bone sample showed that the degree of material symmetry depends strongly on the chosen RVE size [6]. Due to this findings, one wants to determine in the analysis of a complete femur how strong the degree of material symmetry varies throughout the complete bone.

The currently used algorithm is implemented in Co-array Fortran as a general tensor rotation with a full coverage of the 3 dimensional $[\alpha, \vartheta, \phi]$ parameter space with the target function

$$MS(\alpha, \vartheta, \phi) = \left[\sum_{i=1}^{3} \sum_{j=4}^{6} E_{ij}^2 \right] + E_{45}^2 + E_{46}^2 + E_{56}^2 \tag{10}$$

where E_{ij} are the components of $\overline{C}$ expressed in terms of the well known 6×6 matrix formulation.

Although the implemented technique has the advantage that the global minimum of the target function can always be found, it was shown during this study, that it is not feasible to apply the algorithm to the number of RVEs resulting for example from the analysis of a complete human femur. The problem is the walltime which is 150 sec for the current implementation running on 128 Cores of the Cray XT5m installed at HLRS. Since it is assumed that the algorithm can be accelerated by implementing a special version of a rank four tensor rotation as well as using an optimization strategy rather than a complete parameter space coverage it is not considered in the results section.

3 Results

In this section first the measurements done on the single process chain parts for a single RVE are presented along with an estimation of the number of RVEs resulting from the analysis of all regions of interest in a human femur. After that the accumulation of File number, I/O size and Single Core Walltime of the process for the analysis of the estimated number of RVEs is presented followed by resource estimations for the systems currently installed at HLRS.

3.1 Single Sub-Domain

From Table 1 it can be seen, that all process parts together generate approximately 2.75 GB of File I/O during 562 sec runtime. This results in an I/O rate of 4.94 MB/sec which has to handled by one core since all process parts are scalar implementations.

Since the data transformation part after the FE simulations generates redundant data, in Sect. 3.3 only the output I/O size of the Geometry discretization, the result transformation and the calculation of $\overline{C}$ is considered in the calculation of the needed storage size.

Table 1 File number, I/O size and walltime per process part

Process part	Input I/O		Output I/O		Walltime [sec]
	Files	Size [MB]	Files	Size [MB]	
Geometry discretisation	1	0.95	7	18.15	4.1
FE-Simulations	7	18.15	6	940.0	264.0
Result Transformation	6	940.0	24	430.0	290.0
Calculation of $\overline{C}$	24	430.0	1	$0.3 \cdot 10^{-3}$	4
Sum	**38**	**1389.1**	**38**	**1388.15**	**562.1**

3.2 Domain Count

The epiphyseal volume of an average human femur, as shown in Fig. 4, is $\approx 321110\ \text{mm}^3$. The number of resulting RVEs form this volume for the RVE edge lengths 1.2 mm and 1.6 mm is shown in Table 2.

The two given RVE sizes were chosen according to the resolution of currently available clinical CT scanners which reach a minimum isotropic voxel size of approximately 0.4 mm. Since the application of the presented process chain generates an anisotropic elastic material model, 21 material constants have to be determined for each RVE. So the chosen RVE edge lengths of 1.2 mm and 1.6 mm will give 27 or 64 correlation parameters respectively to map the 21 unknowns to.

Table 2 Domain count in dependence of RVE edge length

	1.2 mm	1.6 mm
RVE volume mm^3	1.728	4.096
Epiphyseal Volume mm^3	321110	321110
RVE Number	**185828**	**78396**

3.3 Accumulation over Process

In Table 3 the accumulation of the resource requirements over the complete process chain for one RVE, for the number of RVEs resulting from a RVE edge length of 1.2 mm and for the number of RVEs resulting from a RVE edge length of 1.6 mm are shown.

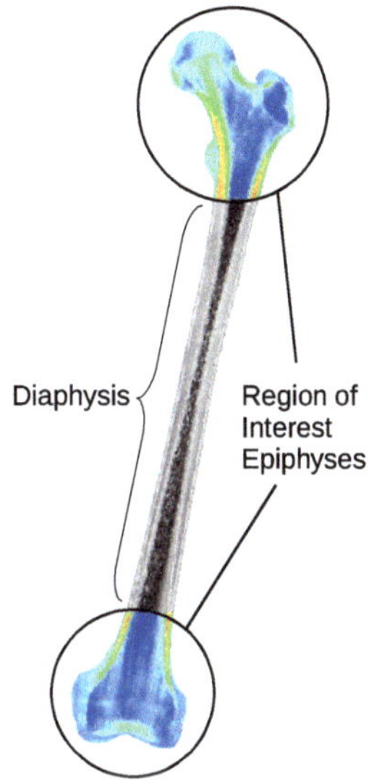

Fig. 4 Regions of interest in a human femur

Table 3 File number, I/O size and walltime—accumulated over process—per core

Domain No.	1	185828	78396
File No.	32	5946496	2508672
I/O [TB]	0.0027	492	208
Storage [TB]	0.0005	93	39
Walltime [s]	562	104435336	44058552
Walltime [h]	0.16	20910	12238

From Table 3 it can be seen, that the process will generate more than an usual load to a file system in terms of file number and storage space. The extrapolation shows on the one hand, that it is necessary to generate a directory tree system to avoid performance penalties due to long file system query times which will come up when a large number of files reside within one directory.

On the other hand it has to be carefully looked at the Storage size of the complete process which exceeds in the case of 1.2 mm RVE edge length the size of a scratch file system which is installed on current medium sized PC-clusters e.g. the HLRS NEC-Nehalem Cluster.

To analyze the resulting walltime and effects when the process chain is executed for many RVEs in parallel the accumulated values from Table 3 are shown in Table 4 for the cases when one of the two HLRS PC-clusters or the Cray XT5m installed at HLRS are completely used to determine anisotropic, elastic material data for RVEs in parallel.

It can be seen form Table 4 that the I/O rate of one core per second neither depends on the number of RVEs nor on the number of cores, since the application

Table 4 I/O and walltime—accumulated over process—parallel execution

System	XT5m		BW-Grid		Nehalem	
Cores	896		3984		5600	
Domain No.	185828	78396	185828	78396	185828	78396
Walltime [h]	32.38	13.66	7.28	3.07	5.18	2.19
I/O / Core [MB/s]	4.94		4.94		4.94	
I/O [MB/s]	4428		19687		27673	

of the process chain is a trivial parallel task and with a value of 4.94 MB/s the I/O rate isn't a problem even if it is accumulated for one of the nowadays used quad core nodes.

If we now take a look at the last row of Table 4 the real problem of the proposed implementation can be seen. I/O rates between 4428 MB/s and 27673 MB/s are way beyond the I/O rates for example a Lustre file system connected via 4x DDR InfiniBand, like the one installed at the HLRS NEC-Nehalem cluster, can make available.

To overcome this problem two solutions can be proposed. One would be to reduce the average I/O load to a reasonable value of e.g. 1500 MB/s which can be made available by nowadays file systems. This can be achieved by reducing the number of used CPU cores to ≈ 300. The other way would be to replace the file system access by an I/O forwarding library like IOFWD[2] and in that way connect the single process parts without having to write to disk at all.

4 Summary & Conclusions

A simulation process chain was presented, which enables the calculation of elastic material constants from micro-CT data. The resource requirements of the four basic parts of the process chain, geometry setup, static simulations, calculations of the effective stiffness and transformation of the effective stiffness were analyzed in terms of CPU time and I/O bandwidth requirements.

The accumulation of the single process parts over the complete process chain and an actual region of interest, like the epiphyseal regions of a human femur, showed that, from the view of walltime requirements, it should be possible to determine the field of anisotropic elastic constants of the region of interest in a human femur within one day.

To reduce the walltime as well as the I/O load the replacement of the commercial FE-Package ABAQUS by a FE-Package with accessible source code is currently under development to make the FE result transformation obsolete. For a further reduction of the file system I/O load the possibilities of the I/O forwarding library IOFWD are currently analyzed.

[2] https://gforge.hlrs.de/projects/sxlinux/

References

1. S.J. Hollister, N. Kikuchi, Computational Mechanics, **10**, (1992), pp. 73–95.
2. B. van Rietbergen, H. Weinans, R. Huiskes and A. Odgaardt, J. Biomechanics, **28** 1, (1995), pp. 69–81.
3. J. Schreiner, C.E. Scheidegger, S. Fleishman, C.T. Silva, EUROGRAPHICS, **25** 3, (2006).
4. M.J. Ciarelli, S.A. Goldstein, J.L. Kuhn, et al., Journal of Orthopaedic Research, **9**, (1991), pp. 674–682.
5. J.Y. Rho, M.C. Hobatho, R.B. Ashman, Medical Engineering and Physics, **17**, (1995), pp. 347–355.
6. R.Schneider, U.Hindenlang, M. Resch, 3rd ANSA & μETA International Conference, September 9–11, (2009), Halkidiki, Greece.

References

Part V
Computational Fluid Dynamics

Downscaling Climate Simulations for Use in Hydrological Modeling of Medium-Sized River Catchments

P. Berg, H.-J. Panitz, G. Schädler, H. Feldmann, Ch. Kottmeier

Abstract To assess a possible future change in flood and drought risks for medium and small-scale river catchments, one needs to have data of a higher spatial and temporal resolution than what is provided by the global climate models. The COSMO-CLM regional climate model has to this purpose been used to downscale a set of global climate simulations to a 7 km horizontal resolution. In order to assess some of the uncertainties involved in near future scenario simulations, several different global simulations are downscaled to produce an ensemble of high resolution data. This will then be used as input to hydrological catchment models to assess future changes in flood risk for three catchments in Germany, within the CEDIM-project "Flood hazard in a changing climate" (Hochwassergefahr durch Klimawandel).

1 Introduction

A set of high resolution climate simulations are carried out on the HLRS facilities in Stuttgart. The main research question to answer is whether there is an elevated risk for floods in medium and small-scale river catchments in Germany. The simulations are performed as a part of the CEDIM-project "Flood hazard in a changing climate" (Hochwassergefahr durch Klimawandel), with a focus on the near future (2021–2050). Three representative catchments have been chosen and a set of hydrological simulations, using different models, will be performed for each of them. The catchments are: Ruhr in the west, Mulde in the east and Ammer in the south. Such small catchments require high quality input data, which must have both spatial and temporal characteristics close to observations.

Peter Berg, Hans-Jürgen Panitz, Gerd Schädler, Hendrik Feldmann, Christoph Kottmeier
Institut für Meteorologie und Klimaforschung, Karlsruhe Institut für Technologie (KIT), Karlsruhe, Germany
e-mail: berg@kit.edu

M. Resch et al. (eds.), *High Performance Computing on Vector Systems 2010*,
DOI 10.1007/978-3-642-11851-7_12,

The COSMO/CLM (CCLM) [1] is used to downscale general circulation model (GCM) simulations. Several GCM simulations are being downscaled in order to assess some of the uncertainties involved in future climate change. One large source of uncertainty is the natural variability, which is comparable to the climate change signal for the next few decades. This uncertainty is addressed by performing downscaling simulations of initial condition perturbed GCM simulations. Here, three different realizations of the ECHAM5-OM GCM [2–7], each initialized in a slightly different natural mode, are used. Another uncertainty arises due to the GCM physics, e.g. details of how cloud processes are simulated. To address this we will perform additional simulations with other GCMs.

In this article, the CCLM model and the downscaling procedure are described in Sect. 2, the performance on the HLRS system is reviewed in Sect. 3, and some results from the validation simulations are presented in Sect. 4. We end with a short discussion in Sect. 5.

2 The CCLM Model

As described in [8] and [9], dynamical downscaling is used to transfer large scale information to the regional scale. Basically, this method is a nesting of the regional climate model (RCM) into large-scale global circulation model (GCM) projections or reanalyzes. This means that the model is initialized once with a state derived from the large scale information and that this information is updated at the lateral boundaries of the regional model domain at regular time intervals.

To validate the model, we perform downscaling simulations with the ERA40 reanalysis data [10]. This data set comes from a global simulation where also observations are assimilated, and it is therefore close to the observations while also filling in the gaps between the observations. The product covers the period 1957–2002 and is a good representation of climate for that period. The downscaling performed with the CCLM can therefore be directly compared to observational records.

For the CEDIM project, the CCLM model version 4.8 is used and the forcing GCM is ECHAM5, which itself, in the control simulations (1971–2000), is forced by anthropogenic emissions [2–4]. The future scenario (2021–2050) uses the A1B SRES [11] scenario [5–7]. Three simulations with different initial conditions are performed for each of the control and future scenario periods.

The horizontal grid sizes of the global data sets are larger than 100 km (about 125 km for ERA40 and 150 km for ECHAM5). To avoid a too large step in the grid sizes, a double nesting technique is used for the dynamical downscaling. In a first step the large scale data are used to drive a CCLM simulation with a horizontal grid size of 0.44 deg (about 50 km at our latitudes). The chosen model domain for this first nest (Nest 1) is shown in Fig. 1 (left). In the West-East direction the domain consists of 118 grid-points, in the South-North direction of 110 grid-points. In the second step the results of this 50 km simulation are used to drive the CCLM simulations with a grid spacing of 0.0625 deg (about 7 km). The corresponding

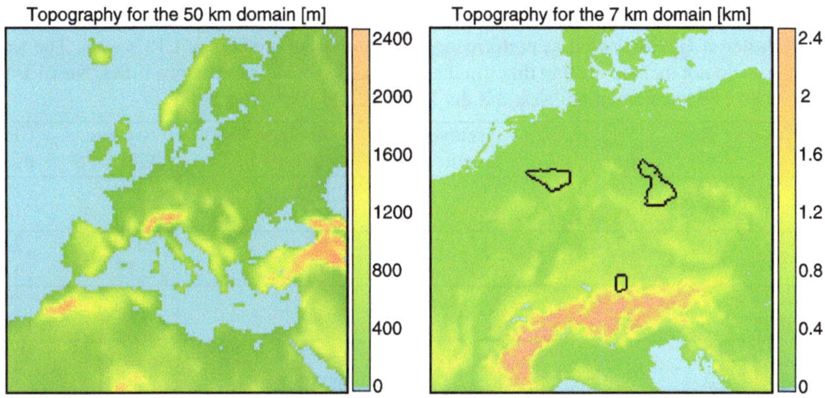

Fig. 1 Illustration of model domain of the Nest 1 (left) with a horizontal resolution of 50 km, and Nest 2 for the CEDIM-simulations (right) with a 7 km resolution. The river catchments of the CEDIM-project are also outlined in the right figure

model domain of this second nest (Nest 2), with 180*200 grid points, is illustrated in Fig. 1 (right).

3 Performance on the HLRS Systems

In a previous project, a total of 16 CCLM simulations have been carried out within the research program "Herausforderung Klimawandel" with focus on the state of Baden-Württemberg ([12, 13]). The simulations within this project are similar to those for the present project. Even though an older version of the CCLM model was used, the difference in performance is negligible. The overall computational needs were rather large with a required wall-clock time of about 198 days and a total CPU-time of about 20 years.

Within the CEDIM project, eight additional simulations have already been performed at the High Performance Computing Center Stuttgart (HLRS) facilities, and another six simulations are currently running or are in the pipeline for the next few months. Due to a larger domain size of Nest 2 the computational needs are even larger than for the Baden-Württemberg project. Performance details for the simulations are summarized in Table 1 for each platform. The bulk of the simulations have been carried out using the NEC SX8 high performance computer, and some are carried out on the SX9 and on the BWGRID cluster.

As can be seen in Table 1, the Nest 2 simulations are more computationally demanding. This is due to a shorter numerical time-step and a larger grid-domain, which has a larger contribution for the newer simulations with a domain almost a factor of two larger than the previous ones. There is a strong increase in performance for the SX9 compared to the SX8, but the BWGRID cluster is also performing well although requiring several times more CPUs.

In order to optimize the numerical code of CCLM a cooperation between HLRS and IMK has been initiated, which is very much appreciated.

Table 1 Wall-clock times needed for CCLM simulations on NEC SX8, NEC SX9 and the BW-GRID cluster at HLRS as well as performance indicators and numbers of CPUs used. The wall-clock times do not include waiting times in the queue and system maintenance times. Nest 1 is the 118*112 grid 50 km simulation, Nest 2 is the 180*200 grid 7 km simulation

Platform	Nest No	Wall-clock time (d)	Average vector length	Average vector ratio (%)	Average MFLOPS	CPUs used
SX8	1	9	112	86	1400	16
SX8	2	28	159	98	3600	16
SX9	1	2.3	117	97	3680	24
SX9	2	8	127	97	3275	32
BWGRID	1	9	–	–	–	80
BWGRID	2	42	–	–	–	168

4 Results

The recent simulations for the CEDIM project show an improved performance, especially when it comes to precipitation. Figure 2 presents some first results of the comparison of the ERA40 reanalysis driven simulations (CE40_7k) at a 7 km resolution for the three river catchments. The fairly large uncertainties in the observational data makes it challenging to validate the RCM simulations. Here we have used three different sources of observations; E-OBS [14] which is originally at a 25 km grid, REGNIE [15] at a 1 km grid, and a 7 km gridded version of a homogenized PIK/DWD station data set [16], which we refer to as "PIK". All three data sets are interpolated to the CCLM 7 km grid for easier comparison. Besides the resolution, the data sets differ due to two main reasons: (i) the number of station observations underlying the data set; REGNIE and PIK have high densities of stations while E-OBS has an order of magnitude less, (ii) undercatch correction; REGNIE is corrected for the problems the precipitation gauges have in catching all precipitation for windy and snowy conditions [17], while the other two data sets are based on uncorrected measurements.

The uncertainty in the observed value from these data sets is evident from Fig. 2. E-OBS shows generally lower monthly precipitation by 10–30% compared to the

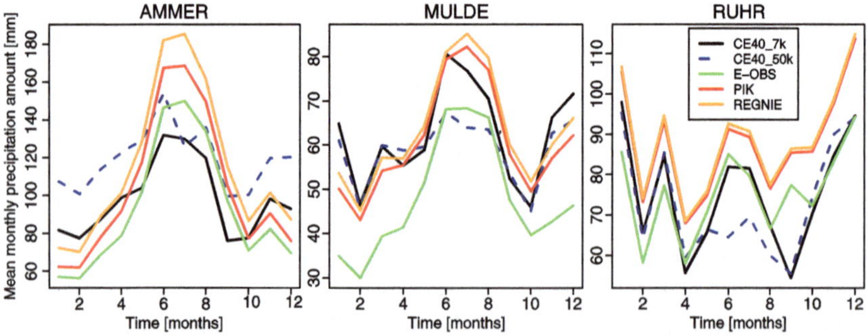

Fig. 2 Mean monthly precipitation sum for 30 years simulation with the CCLM model (black), compared to different observational data sets as explained in the text, for the Ammer (left), Mulde (middle) and Ruhr (right) river catchments

other data sets. PIK and REGNIE are relatively close with REGNIE having slightly higher values, possibly due to the undercatch correction. The seasonal cycle of precipitation in the catchments can be classified into "summer type" for the Ammer and the Mulde, while the Ruhr has more of a "low-mountain type" precipitation characteristic [18].

For the semi-alpine Ammer catchment, Fig. 2 (left) shows that the 7 km simulation is under-estimating the precipitation in summer, while over-estimating in winter. A large part of this is due to the performance over steep terrain in the mountains, which is a general problem for most RCMs even at this high resolution. The performance is better for the Mulde, but for the Ruhr there is a general underestimation of precipitation throughout the year. Note also how the performance for the 7 km simulation is improving on the seasonal cycle as simulated by the model at a 50 km resolution (interpolated to the 7 km grid in the same way as the observations). The improvement is especially large for the summer-time precipitation, where the convection is better reproduced by the high resolution simulation.

The spatial patterns produced by the model are in fairly good agreement with the REGNIE and PIK observations, see Fig. 3. The E-OBS data set once again differs

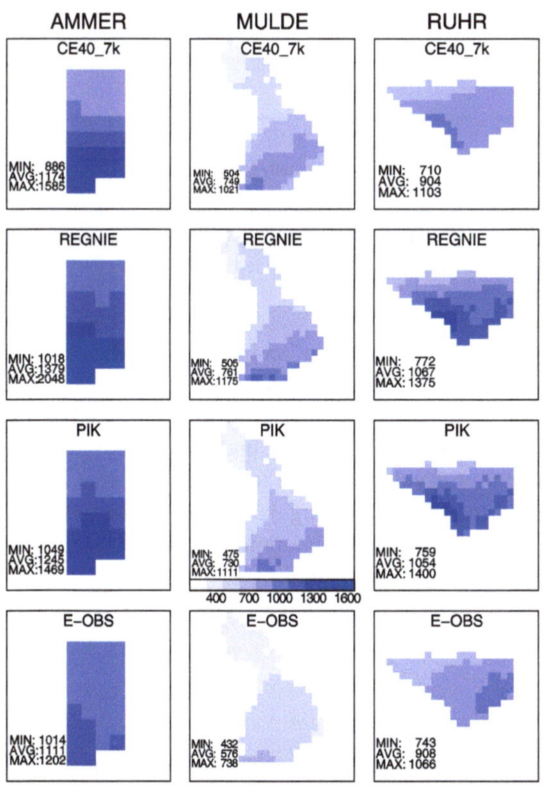

Fig. 3 Mean annual precipitation sum for 30 years simulation with the CCLM model driven by ERA40 (top), compared to different observational data sets as explained in the text, for the Ammer (left), Mulde (middle) and Ruhr (right) river catchments

from the others by both strongly under-estimating the precipitation amount and by not agreeing on the spatial patterns, which is probably due to the much lower density of stations for this data set.

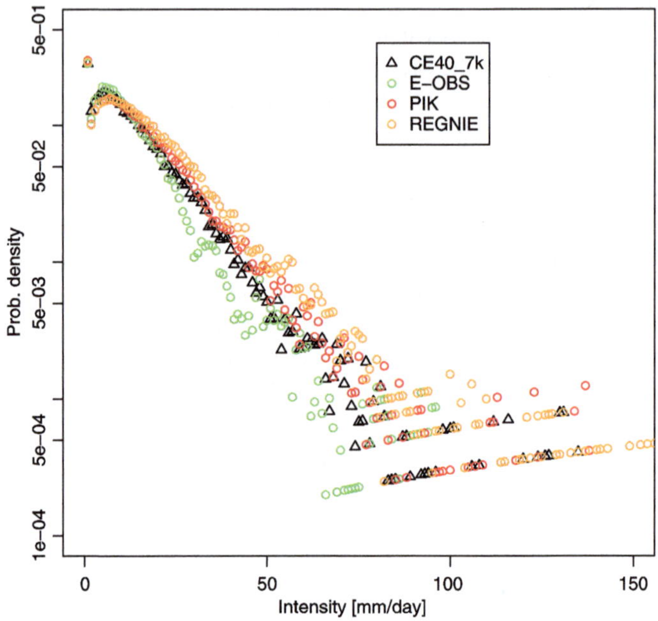

Fig. 4 Probability density function of precipitation intensities of the Ammer catchment for 30 years simulation with the CCLM model driven by ERA40 in comparison to the three observational data sets. Dry days ($< 1\ \text{mm day}^{-1}$) are here included in the PDF at the first position

Figure 4 shows the PDF of precipitation intensity for the Ammer catchment. The REGNIE data set has a slightly more intense distribution than the PIK, which is probably due to the undercatch correction for this data set. The E-OBS data set is over-estimating the lower intensities, and under-estimating the higher intensities. However, the number of dry days is close to that of the other data sets. The CCLM simulation driven by the ERA40 forcing data is very close to the REGNIE and PIK data sets, with only a slight under-estimation for the more extreme intensities. Also the number of dry days is well simulated.

5 Discussion

Simulations of hydrological processes in small river catchments with complex terrain require high spatial detail of the forcing data. The simulations presented above are performing well in the spatial structures, and also fairly well in the absolute amounts. Furthermore, it has been shown that the distributions of precipitation intensities compare well with high resolution observations, something which is very important when assessing extreme events such as floods.

The ensemble aspect of climate modeling is central to assessing probabilities for future changes, in e.g. precipitation. It is important to include several GCM-simulations in any regional assessment of future climate, and the more models used the better the sampling of the uncertainty range, assuming that the models cluster around the "truth". One additional member to the IMK CCLM ensemble will be simulations based on the results of the third GCM generation of the Canadian Center for Climate Modeling and Analysis (CGCM3, [19, 20]) which are presently being prepared.

More information about the CEDIM-project can be found on the web page www.cedim.de.

References

1. Doms, G. and U. Schättler (2002): A description of the nonhydrostatic regional model LM, Part I: Dynamics and Numerics. COSMO Newsletter, 2, 225–235.
2. Röckner, E. (2006a): IPCC-AR4 MPI-ECHAM5_T63L31 MPI-OM_GR1.5L40 20C3M run no.1: atmosphere 6 HOUR values MPImet/MaD Germany. World Data Center for Climate. [doi:10.1594/WDCC/EH5-T63L31_OM-GR1.5L40_20C_1_6H]
3. Röckner, E. (2006b): IPCC-AR4 MPI-ECHAM5_T63L31 MPI-OM_GR1.5L40 20C3M run no.2: atmosphere 6 HOUR values MPImet/MaD Germany. World Data Center for Climate. [doi:10.1594/WDCC/EH5-T63L31_OM-GR1.5L40_20C_2_6H]
4. Röckner, E. (2006c): IPCC-AR4 MPI-ECHAM5_T63L31 MPI-OM_GR1.5L40 20C3M run no.3: atmosphere 6 HOUR values MPImet/MaD Germany. World Data Center for Climate. [doi:10.1594/WDCC/EH5-T63L31_OM-GR1.5L40_20C_3_6H]
5. Röckner, E., Lautenschlager, M., Schneider, H. (2006a): IPCC-AR4 MPI-ECHAM5_T63L31 MPI-OM_GR1.5L40 SRESA1B run no.1: atmosphere 6 HOUR values MPImet/MaD Germany. World Data Center for Climate. [doi:10.1594/WDCC/EH5-T63L31_OM-GR1.5L40_A1B_1_6H]
6. Röckner, E., Lautenschlager, M., Schneider, H. (2006c): IPCC-AR4 MPI-ECHAM5_T63L31 MPI-OM_GR1.5L40 SRESA1B run no.2: atmosphere 6 HOUR values MPImet/MaD Germany. World Data Center for Climate. [doi:10.1594/WDCC/EH5-T63L31_OM-GR1.5L40_A1B_2_6H]
7. Röckner, E., Lautenschlager, M., Schneider, H. (2006b): IPCC-AR4 MPI-ECHAM5_T63L31 MPI-OM_GR1.5L40 SRESA1B run no.3: atmosphere 6 HOUR values MPImet/MaD Germany. World Data Center for Climate. [doi:10.1594/WDCC/EH5-T63L31_OM-GR1.5L40_A1B_3_6H]
8. Meissner, C. and G. Schädler (2007): Modelling the Regional Climate of Southwest Germany: Sensitivity to Simulation Setup. In: High Performance Computing in Science and Engineering '07 [W.E. Nagel, D. Kröner, M. Resch (Eds.)]. ISBN 978-3-540-74738-3, Springer Berlin Heidelberg New York.
9. Meissner, C., G. Schädler, H.-J. Panitz, H. Feldmann, and Ch. Kottmeier (2009): High resolution sensitivity studies with the regional climate model COSMO-CLM. Meteoroligsche Zeitschrift, 18, 543–557, doi:10.1127/0941-2948/20090400.
10. Uppala, M., P.W. Kållberg, A.J. Simmons, U. Andrae, V. Da Costa Bechtold, M. Fiorino, J.K. Gibson, J. Haseler, A. Hernandez, G.A. Kelly, X. Li, K. Onogi, S. Saarinen, N. Sokka, R.P. Allan, E. Andersson, K. Arpe, M.A. Balmaseda, A.C.M. Beljaars, L. Van de Berg, J. Bildlot, N. Bormann, S. Caires, F. Chevallier, A. Dethof, M. Dragosavac, M. Fischer, M. Fuentes,

S. Hagemann, E. Hólm, B.J. Hoskins, L. Isaksen, P.A.E.M. Janssen, R. Jenne, A.P. McNally, J.-F. Mahfouf, J.-J. Morcrette, N.A. Rayner, R.W. Saunders, P. Simon, A. Sterl, K.E. Trenberth, A. Untch, D. Vasiljevic, P. Viterbo, J. Woolen, (2005): The ERA40 re-analysis, Q.J.R Meteorol. Soc., 131, 2961–3012.

11. Nakicenovic, N. et al. (2000). Special Report on Emissions Scenarios: A Special Report of Working Group III of the Intergovernmental Panel on Climate Change, Cambridge University Press, Cambridge, U.K., 599 pp. Available online at: http://www.grida.no/climate/ipcc/emission/index.htm.
12. Panitz, H.-J., G. Schädler, and H. Feldmann (2010): Modelling Regional Climate Change in Southwest Germany, In: High Performance Computing in Science and Engineering '09 [W.E. Nagel, D. Kröner, M. Resch (Eds.)]. ISBN 978-3-642-04664-3, Springer Berlin Heidelberg New York 2010.
13. Panitz, H.-J., G. Schädler, and H. Feldmann (2009): Modelling Regional Climate Change in Southwest Germany. Oral presentation, The 12th Results and Review Workshop of the HLRS, October 8–9, 2009, at the High Performance Computing Center Stuttgart.
14. Haylock, M. R., N. Hofstra, A. M. G. Klein Tank, E. J. Klok, P. D. Jones and M. New (2008): A European daily high-resolution gridded data set of surface temperature and precipitation for 1950–2006. J. Geophys. Res., 113, D20119, doi:10.1029/2008JD010201.
15. Deutscher Wetterdienst (DWD), Germany (2008).
16. Deutscher Wetterdienst (DWD)/Potsdam-Institut für Klimafolgenforschung (PIK), Germany (2008): Meteorologischer Datensatz für Deutschland 1951–2006.
17. Yang, D., E. Elomaa, A. Tuominen, A. Aaltonen, B. Goodison, T. Gunther, V. Golubev, B. Sevruk, H. Madsen, and J. Milkovic (1999): Wind-induced precipitation undercatch of the Hellmann gauges. Nordic Hydrology, 30, 57–80.
18. BMU (2003): Hydrologischer Atlas von Deutschland, Freiburg, ISBN:3-00-005624-6.
19. McFarlane, N.A., J.F. Scinocca, M. Lazare, R. Harvey, D. Verseghy, and J. Li (2005): The CCCma third generation atmospheric general circulation model. CCCma Internal Rep., 25 pp.
20. Scinocca, J. F., N. A. McFarlane, M. Lazare, J. Li, and D. Plummer, 2008: The CCCma third generation AGCM and its extension into the middle atmosphere. Atmos. Chem. and Phys., 8, 7055–7074.

DNS of Rising Bubbles Using VOF and Balanced Force Surface Tension

Hendrik Weking, Jan Schlottke, Markus Boger, Philipp Rauschenberger,
Bernhard Weigand, Claus-Dieter Munz

Abstract The rise behavior of small bubbles in a quiescent environment has been investigated by direct numerical simulation (DNS) using the Volume of Fluid (VOF) method and surface tension modeling based on the balanced force approach. The origin of spurious currents using standard (CSF, CSS) models is shown in detail, emphasis is put on the spatial discretization and the calculation of local curvatures. The effect of the new surface tension model on the resulting rise behavior for different bubble diameters is presented.

1 Introduction

Bubbly flows play an important role in many industrial applications such as fermentation reactors floating or loop reactors. The improvement of the efficiency of these reactors requires a deeper understanding of the rise behavior of gaseous bubbles. Therefore this issue has been the topic of many experimental and numerical studies, e.g. [12, 18, 9]. A wide range of contributions to the subject of the motion of bubbles, drops and particles is contained in the book of Clift, Grace and Weber [2]. Koebe [10] employed the ITLR inhouse code FS3D to investigate the rise of bubbles. The FS3D code is also used in this investigation to simulate rising air bubbles in quiescent water. The treatment of fluid interfaces is a crucial factor for the direct numerical simulation (DNS) of two-phase flow. As the interface represents a

Hendrik Weking, Jan Schlottke, Philipp Rauschenberger, Bernhard Weigand
Institut für Thermodynamik der Luft- und Raumfahrt (ITLR), Universität Stuttgart, Pfaffenwaldring 31, 70569 Stuttgart, Germany
e-mail: hendrik.weking@itlr.uni-stuttgart.de

Markus Boger, Claus-Dieter Munz
Institut für Aerodynamik und Gasdynamik (IAG), Universität Stuttgart, Pfaffenwaldring 21, 70569 Stuttgart, Germany
e-mail: boger@iag.uni-stuttgart.de

M. Resch et al. (eds.), *High Performance Computing on Vector Systems 2010*,
DOI 10.1007/978-3-642-11851-7_13, © Springer-Verlag Berlin Heidelberg 2010

discontinuity separating the two fluids, jump conditions have to be applied. Using the so called one-fluid formulation, one is able to work with a single set of equations for the whole flow domain. The coupling of the fluids at the interface is taken into account by the use of variable material properties. Moreover, the Navier-Stokes equations contain an additional volume force term for the surface tension, which is directly linked to the interface location via a delta function concentrated on the surface. From a physical point of view the surface force is balanced by the pressure jump across the interface. As the volume force is only different from zero at the interface, no additional jump conditions with regard to momentum conservation have to be applied.

In general, surface tension is numerically approximated by a volume force using a continuum model. Applying the widespread continuum surface force (CSF, [1]) or continuum surface stress (CSS, [11]) model, many DNS codes suffer from unphysical parasitic currents in the vicinity of the surface.

In particular, the occurrence of spurious currents prevents the investigation of small bubbles with diameters $d_e < 2$ mm. For these cases the numerically induced, parasitic accelerations increase the magnitude of the physical, gravity driven accelerations and thus no benefit can be extracted from such simulations.

In order to overcome this issue, many efforts have been taken by various researchers. For two dimensional computations Meier et al. [13] proposed a method to determine curvature more accurately using an estimator function, tuned with a least-squares-fit against precomputed reference data. An approximation of the surface tension based on spline interpolants was presented by Ginzburg and Wittum [5]. For 3D calculations, Renardy and Renardy [16] developed the PROST algorithm (parabolic reconstruction of surface tension) and Jafari et al. [8] presented the PCIL method (pressure calculation based on the interface location) where the pressure forces at the interfacial cell faces are calculated according to the pressure imposed by each fluid on the portion of the cell face that is occupied by that fluid. This list is far from being complete as there are many other approaches aiming at the reduction of parasitic currents.

The method used for the calculations presented in this paper is based on the balanced force approach by François et al. [4]. Starting from the original CSF model of Brackbill et al. [1] only minor changes are needed, mainly linked to curvature calculation which we choose to be evaluated according to Popinet [15].

The outline of the paper is as follows. We firstly introduce the governing equations for the DNS of two-phase flows, as they are used in the framework of the FS3D code. The following section is dedicated to the calculation of surface tension. In this context the origin of parasitic currents is discussed. This includes on the one hand the correct discretization of the surface tension terms in order to have a so called balanced-force algorithm. On the other hand, the issue of curvature estimation has also to be taken into account and for that we are using a height function method. The results section comprises various calculations on the rise behavior of small bubbles, comparing the results of the two standard models (CSF, CSS) and the new surface tension model (CSF-BHF).

2 Governing Equations

2.1 Continuity and Navier-Stokes Equations

The incompressible two phase flow is described by the continuity equation and the Navier-Stokes equations. Using the one-fluid formulation, the continuity equation and the momentum equations have the following form

$$\nabla \cdot \mathbf{u} = 0, \tag{1}$$

$$\frac{\partial(\rho \mathbf{u})}{\partial t} + \nabla \cdot (\rho \mathbf{u}) \otimes \mathbf{u} = -\nabla p + \rho \mathbf{k} + \nabla \cdot \mu [(\nabla \mathbf{u} + (\nabla \mathbf{u})^T)] + \mathbf{f}_\gamma \delta_S. \tag{2}$$

Here $\mathbf{u}$ represents the velocity, ρ denotes the density, p is the pressure, t is the time, μ is the dynamic viscosity, $\rho\mathbf{k}$ takes into account body forces, the term $\mathbf{f}_\gamma$ is a body force that represents the influence of surface tension in the vicinity of the interface and δ_S is a delta function concentrated on the surface.

These equations resemble the conservation equations for one phase flows with the exception of the appearance of the volume force $\mathbf{f}_\gamma$ on the right-hand side of the momentum equations in Eq. 2. This approach is called the one-fluid formulation and it allows the use of a single set of equations in the whole flow domain. In the absence of mass transfer, there is no need of establishing any additional jump conditions at the fluid interface as they are implicitly taken into account by this formulation. The coupling of the two fluids is provided by variable material properties ρ and μ. They are chosen according to the fluid occupying a grid cell. At the interface the volume force $\mathbf{f}_\gamma$ comes into play for taking into account the effects of surface tension.

2.2 Interface Tracking by the VOF Method

The interface tracking is based on the volume of fluid (VOF) method developed by Hirt et al. [7]. For the representation of the phases, an additional variable f is introduced

$$f = \begin{cases} 0 & \text{in the gaseous phase,} \\ 0 < f < 1 & \text{in cells containing a part of the interface,} \\ 1 & \text{in the liquid phase,} \end{cases} \tag{3}$$

which represents the volume fraction of the liquid phase. A two dimensional example is given in Fig. 1. According to the one-fluid formulation, the above mentioned change in material properties is based on the volume fraction variable f

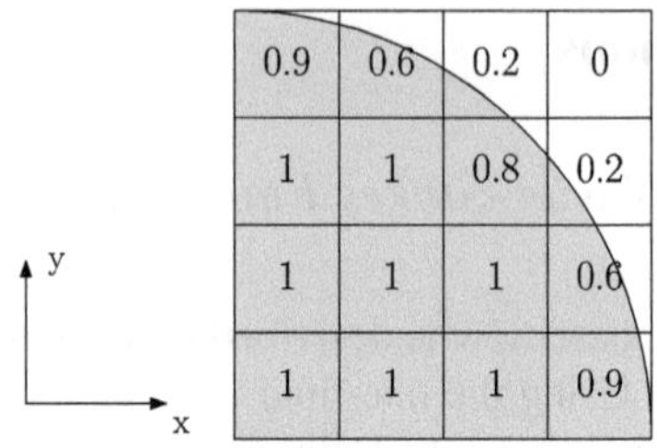

Fig. 1 Fluid representation by the volume fraction f variable

$$\rho(\mathbf{x},t) = \rho_g + (\rho_l - \rho_g) \cdot f(\mathbf{x},t), \tag{4}$$
$$\mu(\mathbf{x},t) = \mu_g + (\mu_l - \mu_g) \cdot f(\mathbf{x},t), \tag{5}$$

where the indices l and g denote the liquid and the gaseous phase, respectively.

The movement of the interface is tracked via the transport of the volume fraction f by

$$\frac{\partial f}{\partial t} + \nabla \cdot (f\mathbf{u}) = 0. \tag{6}$$

While advecting the volume fraction f, one is interested in keeping the interface sharply resolved. Therefore at every time step the interface is reconstructed based on a PLIC (piecewise linear interface calculation) algorithm (see e.g. [17]). This geometrical reconstruction is used to determine the liquid and gaseous fluxes across the cell faces and it prevents the interface from being smeared across several grid cells during the advection.

2.3 Moving Frame of Reference

A moving frame of reference is employed to investigate the bubble motion over a larger time scale and thus over a longer rising path without a large computational domain and the resulting computational effort as presented in [19]. This technique induces a counterflow in the domain which is adapted dynamically over time. Therefore the motion of the center of mass of the bubble is considered as a damped oscillation of a point mass around its initial position. Depending on the displacement and the velocity of the center of mass of the gaseous phase a correction body force $\rho\mathbf{k}_c$ is imposed to the flow field by adding it to the right-hand side of Eq. 2, inducing the counter flow in the computational domain. This keeps the bubble close to its initial position, or from the view of a resting observer, moves the whole domain along with the bubble. This correction force is computed for all three spatial directions in every timestep by

$$\mathbf{k}_c = 2\omega\frac{\mathbf{P}_g}{m^*} + c_2\omega^2\Delta\mathbf{x}_g, \tag{7}$$

with the momentum of the bubble $\mathbf{P}_g$ divided by the bubble virtual mass $m^* = m_g(\rho_f/\rho_g)$ and the displacement vector $\Delta\mathbf{x}_g$. The momentum and the displacement vector refer to the moving frame of reference. To avoid a build up in the displacement the value of c_2 is switched from 1 to 0 if the center of mass is approaching the initial position. The angular frequency governing the hardness of the damping and thus the magnitude of the oscillation around the initial position is chosen to be

$$\omega = c_1\sqrt{\frac{\rho_f}{\rho_g}}\pi\omega_{res} = c_1\sqrt{\frac{\rho_f}{\rho_g}}\pi\sqrt{\frac{12\sigma}{d_e^3\rho_f}}, \tag{8}$$

where ω_{res} is the first resonance mode of the bubble surface. ω_{res} is multiplied by π to avoid that ω becomes a higher mode of that surface resonance frequency and thus prevent a build up of surface oscillations which could destroy the free surface. A value of $c_1 = 10$ was chosen in this study. Note, that this formulation does not fix the bubble at the start position but allows a displacement.

3 Surface Tension

At an interface separating two immiscible fluids a pressure jump

$$p_2 - p_1 \equiv \Delta p = \sigma\kappa, \tag{9}$$

appears according to the Young-Laplace equation. Here σ is the surface tension coefficient and κ represents the curvature. Hence, surface tension is directly proportional to the curvature κ. This jump in pressure has to be considered when simulating two-phase flows. Therefore, different models have been developed in order to include surface tension at the interface. As it is obvious from Eq. 2, surface tension is accounted for as a volume force in the momentum equations. The corresponding force $\mathbf{f}_\gamma$ on the right-hand side of the equation is only present at the interface while it vanishes in grid cells away from the interface.

3.1 The Continuum Surface Force (CSF) Model

The basic idea of the continuum surface force (CSF) model introduced by Brackbill et al. [1] can be described as follows. Instead of considering the fluid interface as a sharp discontinuity, one supposes a smooth transition from one fluid to another. The interface is considered to have a finite thickness of $O(h)$, corresponding to the smallest length scale h resolvable by the computational mesh. Consequently, surface tension is also considered to be of continuous nature and it acts everywhere within the transition region. Brackbill and coworkers propose to calculate $\mathbf{f}_\gamma$ by

$$\mathbf{f}_\gamma = \sigma\kappa\nabla f. \tag{10}$$

This corresponds to a dispersion of the surface tension across the transition region, using the gradient of the volume fraction variable f to weight the dispersed volume force. Implementing this approach in a CFD code, one has to consider the

1. spatial discretization of Eq. 10,
2. estimation of curvature κ,

in order to prevent parasitic currents. The following sections will provide a discussion on both aspects.

3.2 Balanced-Force Algorithm

Looking at the spatial discretization, the variables are stored on a staggered grid arrangement according to [6]. On such a MAC (marker-and-cell) grid, the scalar variables (f, p) are stored at the cell centers, while the velocities are stored at the centers of the cell faces. In order to guarantee an accurate, balanced-force discretization, according to [4], the surface tension terms (cf. Eq. 10) have to be calculated at the center of the cell faces. Furthermore, it is of crucial importance that pressure and surface tension are discretized in the same way.

Moreover, special care has to be taken in order to evaluate the gradient ∇f. For the original CSF method, according to [1], the evaluation of the gradient is performed on a stencil of 18 cells for 3D calculations. For explanation purposes Fig. 2 illustrates a two dimensional example. Here, the gradients at the cell face centers of row (j) are given by the corresponding values in red. Taking the face at $(i+1/2)$ as an example, the gradient $\nabla f_{x_{i+1/2,j}}$ is based on the gradients in the rows $(j-1),(j),(j+1)$ that are calculated on the basis of the cells adjacent to the face $(i+1/2)$ respectively, e.g. $\tilde{\nabla} f_{x_{i+1/2,j}} = \frac{f_{(i+1,j)}-f_{(i,j)}}{\Delta x}$ for row (j). Here, $\tilde{\nabla}$ designates the local gradient with respect to the cell face. Afterwards, the gradient at the position $x_{i+1/2,j}$ is obtained as

$$\nabla f_{x_{i+1/2,j}} = \frac{1}{4}\left(\tilde{\nabla} f_{x_{i+1/2,j-1}} + 2\tilde{\nabla} f_{x_{i+1/2,j}} + \tilde{\nabla} f_{x_{i+1/2,j+1}}\right). \tag{11}$$

This leads to a total stencil for cell (i, j) that is surrounded by the dashed lines in Fig. 2. It is obvious that the above discretization implies a coupling of the rows $(j-1),(j),(j+1)$ for the surface tension calculation via the evaluation of the gradients. Having a closer look at the discretization of the Poisson equation used for the calculation of p, in x direction for the cell (i, j), one finds

$$\frac{1}{\Delta x}\left(\frac{p_{(i+1,j)} - p_{(i,j)}}{\rho_{(i+1/2,j)}} - \frac{p_{(i,j)} - p_{(i-1,j)}}{\rho_{(i-1/2,j)}}\right) = \frac{\tilde{\mathbf{u}}_{(i+1/2,j)} - \tilde{\mathbf{u}}_{(i-1/2,j)}}{\Delta t}. \tag{12}$$

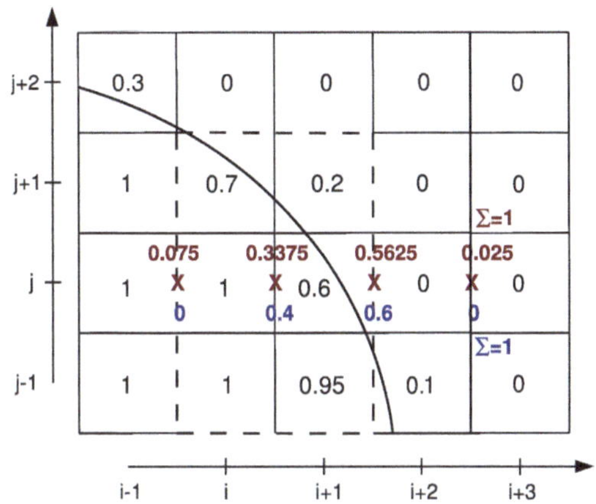

Fig. 2 Evaluation of ∇f based on a stencil of 6 cells (red) and on the direct neighbors of the cell face (blue). The box surrounded by the dashed lines marks the 6 cell stencil for the cell face $(i+1/2, j)$

Here it is clearly visible that all pressure values are taken from row (j). As for the above mentioned discretization of the surface tension used to evaluate the velocity $\tilde{\mathbf{u}}$, the rows $(j-1), (j+1)$ are also included via the gradient evaluation. We found this coupling to be one of the causes for the parasitic currents using the CSF implementation in FS3D. Therefore we changed the gradient evaluation to a more local formulation only taking into account direct neighbors of the cell faces. This leads for the cell face $(i+1/2, j)$ to

$$\nabla f_{x_{i+1/2,j}} = \tilde{\nabla} f_{x_{i+1/2,j}} = \frac{f_{(i+1,j)} - f_{(i,j)}}{\Delta x}. \tag{13}$$

Returning to Fig. 2, the two approaches can be compared directly. According to the two methods, the gradients are given for the previous method (red) and the local approach (blue) in row (j). While the transition from one fluid to the other is spread over four cell faces with the previous approach, the local approach only uses two cell faces to disperse the jump in pressure.

In 3D, the gradients for the different spatial directions are evaluated in an analogous way, only taking into account direct neighbors of the respective cell face.

Besides the discretization of the surface tension force (Eq. 10), the correct estimation of surface curvature is very important and shall be discussed in the following section.

3.3 Curvature Estimation

In the context of the VOF method, difficulties in determining topology information like normal vectors and curvature are due to the discrete nature of the volume fractions. Contrary to a level-set function, the VOF field is not smooth and thus the correct curvature as the second derivative is hard to obtain. Direct calculation of curvature from the given volume fractions leads to high frequency (order of the grid) errors (aliasing error, [3]). This is why most CSF implementations do a smoothing of the VOF field prior to the calculation of curvature. In addition, this error does not vanish with grid refinement.

For the method presented here, the curvature is calculated based on a height function approach, coupled to a local paraboloid fitting if the grid resolution is not sufficient. The procedure is inspired by the article of Popinet [15]. The height function approach is a geometrical one. The stencil used to determine the local height is not fixed (as used by e.g. [4]), but adapts itself to capture the local topology in an optimal way.

As already mentioned, grid resolution does not always permit the determination of curvature via the height function approach. This is the case for highly curved topology, e.g. during breakup, or in case of very coarse grids. Local curvature is then determined by fitting a paraboloid to known points on the surface. These points can either originate from local heights or the barycenters of the reconstructed surface (using PLIC) are used for this purpose. Once the paraboloid, given by

$$f(a_i, \mathbf{x}) = a_0 x^2 + a_1 y^2 + a_2 xy + a_3 x + a_4 y + a_5, \tag{14}$$

is fitted (via least squares fit) to these points, the curvature can easily be calculated as the second derivative,

$$\kappa = 2\frac{a_0(1+a_4^2) + a_1(1+a_3^2) - a_2 a_3 a_4}{(1+a_3^2+a_4^2)^{3/2}}. \tag{15}$$

4 Numerical Setup

Employing the moving frame of reference it is not necessary to discretize the entire rising path. The computational domain can be reduced to a small frame around the bubble, as shown in Fig. 3. On the upper side of the 3D computational domain a uniform inflow condition is employed. On all other boundaries, including the outflow, a continuous (Neumann) condition is used. To oppress undesired back flow a damping zone is placed in front of the outflow boundary. The gravity pointing towards the outflow boundary leads to a strong acceleration of the liquid, which 'falls' out of the domain, taking the bubble with it. This can be avoided by replacing gravity by buoyancy in the momentum equation. In the presented study a variety of bubble diameters were investigated. Therefore the extent of the computational domain varies depending on the initial bubble diameter. The computational domain has an overall length of 20 bubble diameters in the direction of the main flow and 10 bubble diameters between the lateral boundaries. The resolution is chosen to be 512 cells in main flow direction and 256 cells in the lateral direction, which makes about $33 \cdot 10^6$

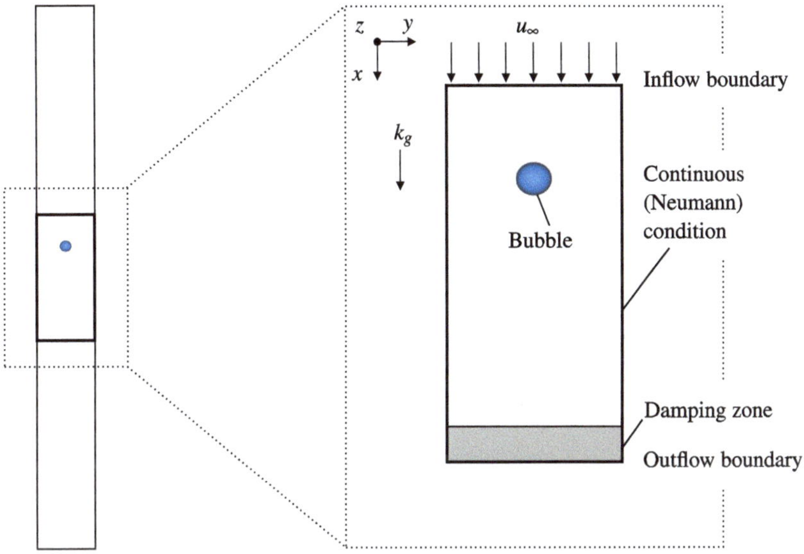

Fig. 3 Numerical setup and coordinate system

computational cells in the whole domain. At this resolution the equivalent bubble diameter is resolved by 25.6 cells.

In all cases the air bubble is initialized as a stationary sphere surrounded by quiescent water at a distance of 5 bubble diameters away from the inflow and the lateral boundaries.

5 Results: Rise Behavior of Bubbles

5.1 Reduction of Spurious Currents

The simulation of small bubbles with an equivalent diameter $d_e < 2$ mm leads to unphysical results using CSF and CSS models. In these cases the spurious currents induced by the high curvature of the interface are no longer negligible. Figure 4 shows the velocity field in a cut plane through the bubble center for $d_e = 1$ mm (left) and $d_e = 2$ mm (right) computed by the CSF, CSS and the new CSF-BHF model. For both bubble diameters the results using the CSF model are strongly influenced by spurious currents. The velocity field computed using the CSS model is apparently free from disturbances caused by spurious currents in case of the 2 mm bubble whereas for the 1 mm case such disturbances appear at the top and the bottom of the bubble. In contrast, the velocity field obtained using the new model is free from spurious currents for both bubble diameters.

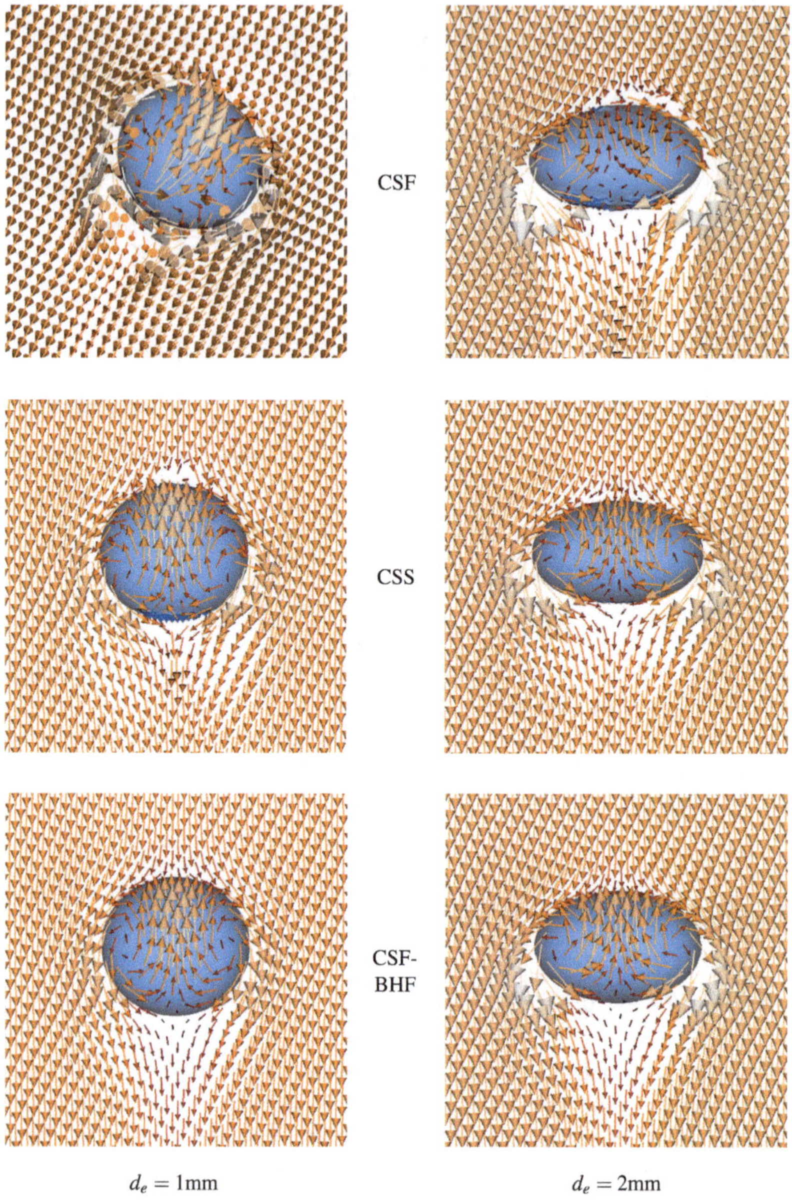

Fig. 4 Velocity field using CSF, CSS and CSF-BHF model for $d_e = 1$ mm (left) and $d_e = 2$ mm (right)

The corresponding development of the rise velocity is plotted in Fig. 5. For the 1 mm case (left) both, the CSF and CSS model, fail to predict a smooth trend due to spurious currents in contrast to the CSF-BHF model. Although the mean rise velocity computed using the CSS model is in the same magnitude as the velocity

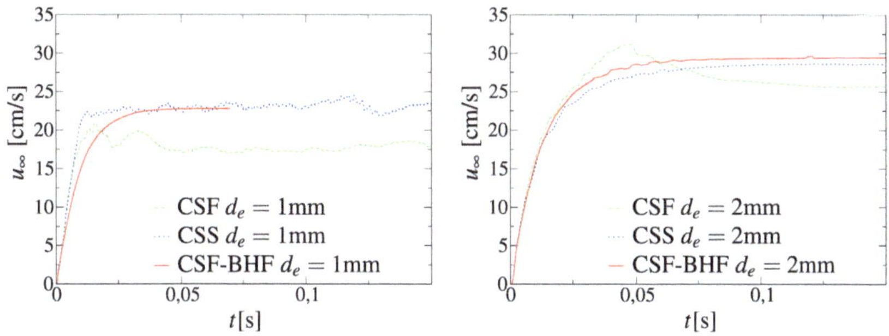

Fig. 5 Comparison of rise velocity using CSF, CSS and CSF-BHF model for $d_e = 1$ mm (left) and $d_e = 2$ mm (right)

gained by the CSF-BHF model, strong oscillations occur due to the nondirectional disturbances of the spurious currents. To demonstrate the influence of the spurious currents near the free surface the ration of the instantaneous interface area to the initial area of the spherical bubble A/A_0, which is the degree of deformation of the bubble, is shown in Fig. 6 (left).

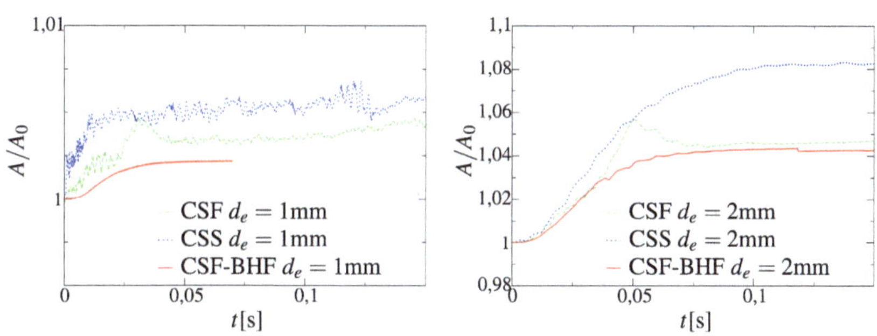

Fig. 6 Comparison of interface area using CSF, CSS and new CSF-BHF model for $d_e = 1$ mm (left) and $d_e = 2$ mm (right)

Only the new model is capable to reproduce a smooth deformation from the sphere to an elliptic shape as the bubble rises. In case of a 2 mm bubble (Fig. 5 right) the CSS and the new CSF-BHF model produce comparable results for the instantaneous rise velocity whereas the rise velocity computed using the CSF model first exceeds the velocity predicted by the other models and decreases to a lower level after this peak. The same peak is found regarding the instantaneous deformation in Fig. 6 (right).

5.2 Terminal Rise Velocity

At the start of the simulation the bubbles are initialized at zero velocity. After a short phase of acceleration, they reach a maximum rise velocity. In this phase the rising trajectory of the bubble stays rectilinear. Depending on its equivalent diameter the motion of the bubble can then turn into a zigzagging or spiraling path. In these cases, the terminal rise velocity is not the maximum velocity of the rectilinear part but the mean value of the 3D motion part. In the cases investigating 3D motion only occurred for $d_e > 3$ mm.

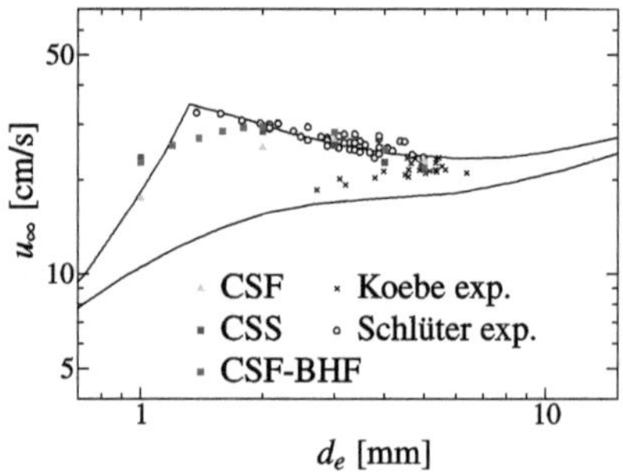

Fig. 7 Comparison of terminal rise velocity with experimental data

The results of the terminal rise velocity are plotted over the equivalent diameter in Fig. 7. The lines in the plot are taken from Clift et al. [2] and represent the terminal rise velocity for air bubbles in pure water (upper line) and in conterminated water (lower line). Regarding the terminal rise velocity the results of all surface tension models fit well with the experimental data measured by Koebe [9] and Schlüter [18] for $d_e \geq 3$ mm. In the region below 2 mm the new surface tension model underestimates the terminal rise velocity, the results are still between the two lines by Clift et al. [2]. The only exception is the 1 mm case where the velocity calculated is too high. Note, that the 1 mm case of the CSS model and the 1 mm and 2 mm cases of the CSF model are affected by spurious currents and are only plotted for the sake of completeness.

5.3 Bubble Shape

The deformation of a bubble is measured by the ratio of the bubble height to its width H/W after reaching the terminal velocity. Obviously, this can only be done when the bubbles develop to a rotational symmetric oblate shape, which is the case for a rectilinear trajectory. In the cases where a 3D rising path occurs, the bubbles develop an arbitrary time dependent shape. The deformation of the bubbles computed using the new model with an equivalent diameter $d_e \leq 2$ mm is presented in Fig. 8 as a function of the Weber number

$$We = \frac{\rho_l d_e u_\infty^2}{\sigma}. \tag{16}$$

The Weber number is the ratio of the kinetic energy to the causing distortion to the surface energy available to resist it. The computed terminal rise velocity shown in Fig. 7 is chosen as the reference velocity u_∞. Figure 8 shows that for small equivalent diameters the simulated deformations match the correlation of Moore [14] for small distortions

$$H/W = \frac{1}{1 + \frac{9}{64}We + O(We^2)}. \tag{17}$$

For larger deformations and thus larger Weber numbers the numerical results deviate from that correlation.

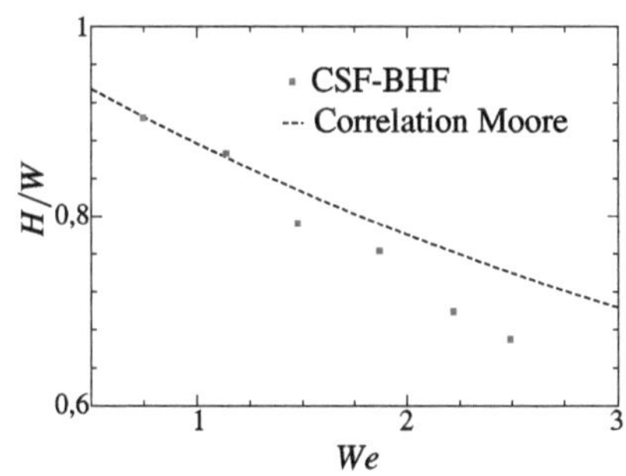

Fig. 8 Bubble deformation H/W for over Weber number computed with CSF-BHF

6 Conclusion

DNS of the rise behavior of small bubbles in a quiescent liquid using the volume of fluid (VOF) method has been presented. The use of an advanced surface tension model based on the balanced force approach and on the calculation of local curvature via height functions enables the investigation of bubbles with diameters smaller than 3 mm where unphysical spurious currents normally dominate the solution when using standard models (CSF, CSS). The predicted rise velocities and bubble deformation fit the experimental data and the correlations from literature. This approach demonstrates that DNS calculations can be used as 'numerical experiments' in order to understand detailed complicated two phase flow situations.

Acknowledgements The authors would like to thank the German Research Foundation (DFG) for financial support of the project within the Cluster of Excellence in Simulation Technology (EXC 310/1) at the University of Stuttgart and the PAK 119. Additionally the authors greatly appreciate the *High Performance Computing Center Stuttgart* (HLRS) for support and computational time on the NEC-SX8 platform under the Grant No. FS3D/11142.

References

1. Brackbill, J.U., Kothe, D.B., Zemnach, C.: A continuum method for modeling surface tension. J. Comput. Phys. **100**, 335–354 (1992)
2. Clift, R., Grace, J.R., Weber, M.E.: Bubbles, Drops, and Particles. Dover Publications Inc., Mineola, New York, USA (2005)
3. Cummins, S.J., Francois,M.M., Kothe, D.B.: Estimating curvature from volume fractions. Comput. Struct. **83**, 425–434 (2005)
4. Francois, M.M., Cummins, S.J., Dendy, E.D., Kothe, D.B., Sicilian, J.M., Williams, M.W.: A balanced-force algorithm for continuous and sharp interfacial surface tension models within a volume tracking framework. J. Comput. Phys. **213**, 141–173 (2006)
5. Ginzburg, I., Wittum, G.: Two-Phase Flows on Interface Refined Grids Modeled with VOF, Staggered Finite Volumes, and Spline Interpolants. J. Comput. Phys. **166**, 302–335 (2001)
6. Harlow, F.H., Welch, J.E.: Numerical Calculation of Time-Dependent Viscous Incompressible Flow of Fluid with Free Surface. Phys. Fluids **8**, 2182–2189 (1965)
7. Hirt, C.W., Nichols, B.D.: Volume of fluid (VOF) method for the dynamics of free boundaries. J. Comput. Phys. **39**, 201–225 (1981)
8. Jafari, A., Shirani, E., Ashgriz, N.: An improved three-dimensional model for interface pressure calculations in free-surface flows. Int. J. Comput. Fluid Dyn. **21**, 87–97 (2007)
9. Koebe, M.: Numerische Simulation aufsteigender Blasen mit und ohne Stoffaustausch mittels der Volume of Fluid (VOF) Methode. PhD thesis, Lehrstuhl für Technische Chemie und Chemische Verfahrenstechnik, Universität Paderborn, Germany, (2004)
10. Koebe, M., Bothe, D., Prüss, J., Warnecke, H.J.: 3D Direct Numerical Simulation of Air Bubbles in Water at high Reynolds numbers. Proceedings of ASME FEDSM'02 (2002)
11. Lafaurie, B., Nardone, C., Scardovelli, R., Zaleski, S., Zanetti, G.: Modelling Merging and Fragmentation in Multiphase Flows with SURFER. J. Comput. Phys. **113**, 134–147 (1994)
12. Maxworthy, T., Gnann, C., Kürten, M., Durst, F.: Experiments on the rising of air bubbles in clean viscous liquids. J. Fluid Mech. **321**, 421–411 (1996)
13. Meier, M., Yadigaroglu, G., Smith, B.L.: A novel technique for including surface tension in PLIC-VOF methods. Eur. J. Mech. B. Fluids **21**, 61–73 (2002)
14. Moore, D.W.: The velocity of rise of distorted gas bubbles in a liquid of small viscosity. J. Fluid Mech. **23**, 749–766 (1965)
15. Popinet, S.: An accurate adaptive solver for surface-tension-driven interfacial flows. J. Comput. Phys. **228**, 5838–5866 (2009)
16. Renardy, Y., Renardy, M.: PROST: A Parabolic Reconstruction of Surface Tension for the Volume-of-Fluid Method. J. Comput. Phys. **183**, 400–421 (2002)
17. Rider, W.J., Kothe, D.B.: Reconstructing Volume Tracking. J. Comput. Phys. **141**, 112–152 (1998)
18. Schlüter, M.: Blasenbewegung in praxisrelevanten Zweiphasenströmungen. PhD thesis, Institut für Umweltverfahrenstechnik, Universität Bremen, Germany, (2002)
19. Weking, H., Huber, C., Weigand, B.: Direct Numerical Simulation of Single Gaseous Bubbles in Viscous Liquids. High Performance Computing in Science and Engineering '09. Springer, Berlin Heidelberg New York (2009)

Large-Eddy Simulation of Double-Row Compound-Angle Film-Cooling: Computational Aspects

Lars Gräf, Leonhard Kleiser

Abstract Film-cooling is an important technique allowing to increase the thermal efficiency of gas turbines. By blowing cool air through an array of small holes in the turbine blades a thin fluid film is set up shielding the blades from the hot gas arriving from the combustion chamber.

This work presents computational aspects of a Large-Eddy Simulation of a particular film-cooling configuration known to provide a high level of effectiveness. The simulation employs the block-structured finite-volume Navier-Stokes code NSMB for compressible flows including the Approximate Deconvolution Model for subgrid turbulence modeling, the Synthetic-Eddy Method for turbulent inflow generation and reflection-reducing outflow boundary conditions. The performance of principal routines is analyzed first for a sequential simulation of a canonical flat plate turbulent boundary layer, showing a high efficiency above 35% of the most time-consuming routines on a NEC SX-8. A strong scaling test of the film-cooling setup shows reasonable parallel speedup up to 32 processors of an SX-8. For both cases, with the new architecture SX-9 a lower relative performance is achieved compared to the SX-8.

1 Introduction

Rising the turbine inlet temperature of gas turbines increases the efficiency but is limited by the turbine blade material properties. Cooling the blades or even protecting them from the hot gases allows for higher inlet temperatures. However, taking the coolant from the compressor as is usually done reduces the mass-flux available for combustion, typically by about one fifth, and hence the overall efficiency of the

Lars Gräf, Leonhard Kleiser
Institute of Fluid Dynamics, ETH Zurich, Sonneggstrasse 3, 8092 Zurich, Switzerland
e-mail: graef@ifd.mavt.ethz.ch, kleiser@ifd.mavt.ethz.ch

M. Resch et al. (eds.), *High Performance Computing on Vector Systems 2010*,
DOI 10.1007/978-3-642-11851-7_14,

gas turbine is reduced. Consequently, the aim is to obtain a high level of cooling at a low coolant mass-flux.

Amongst other techniques [6], film cooling is used to reduce the effective blade temperatures. In order to study basic properties of the cooling arrangement, usually blade curvature and rotational effects are neglected and the flow along a flat plate is considered. In gas-turbine film cooling [4], coolant air blown from the blade interior through discrete holes builds up a coolant film shielding the blade surfaces from the hot oncoming gas, cf. Fig. 1. In a simple case, circular holes are arranged in a spanwise row, cf. Fig. 1c, and are inclined towards the plate and oriented with the principal flow, cf. Fig. 1a.

The simple-angle injection of coolant into the hot boundary layer is known to develop a counter-rotating vortex-pair, termed kidney-vortex. Due to the rotational direction of the vortices, cf. Fig. 1b, hot air is fed towards the plate and the vortex pair tends to lift off from the plate [7]. If the hole additionally has a spanwise yaw angle $\beta \neq 0$, cf. Fig. 1e, f, one single, dominant vortex is known to establish [9]. By combining two subsequent rows of compound-angle holes with opposite spanwise tilt, Fig. 1d, f, an anti-kidney-vortex can be composed with reversed rotational direction, Fig. 1e. Experiments [1] show that this combined vortex is beneficial for the film-cooling effectiveness.

Apparently for the first time, [5] presents a Large-Eddy Simulation (LES) of an anti-kidney film-cooling flow and focuses on the simulation setup as well as the validation against theoretical, numerical and experimental data. The present work reports on computational aspects of the same simulations. Section 2 describes the simulation methods used. The performance of individual code sections is detailed in Sect. 3.1 with the aid of case 1, defined in Fig. 2. The overall performance of case 2 and the flow field are described in Sect. 3.2 and 3.3, respectively. Finally, Sect. 4 summarizes the main findings.

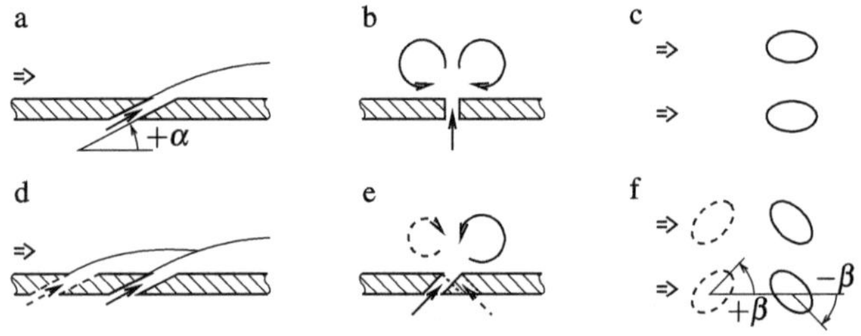

Fig. 1 Film-cooling configurations including flow orientations for a turbine blade simplified as a flat plate. Top: single row, simple angle; bottom: double-row compound-angle injection. Left: side view; center: downstream view; right: top view. Double arrows: hot gas; Dashed: first of two rows. Dominant vortices are indicated in b and e

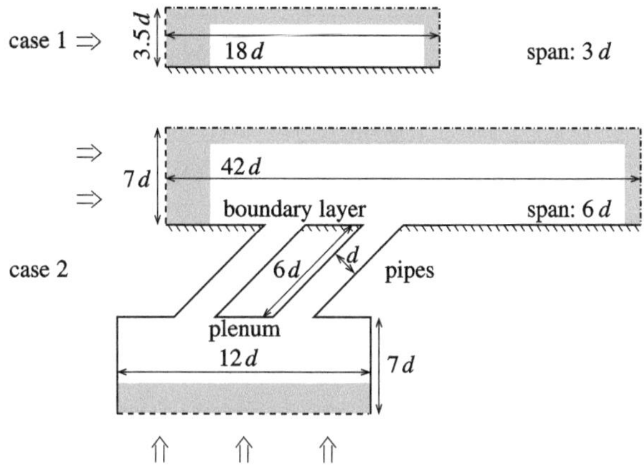

Fig. 2 Computational domains of the two simulation cases. Solid: wall; dashed: inflow boundary; dash-dotted: outflow boundary; shaded: sponge region. (Sketches are not drawn to scale)

2 Simulation Methods

2.1 *Code and Governing Equations*

A detailed description of the simulation setup and parameters is given in [5]. The structured, multi-block, cell-centered, finite-volume simulation code NSMB [15] is used for computation. For the code to run in parallel, the computational domain is decomposed into multiple blocks (the block-grid can be unstructured) and the data exchange between the blocks is done using MPI. Since the maximum centered stencil width is five, each surface of those blocks is extended by two layers of ghost cells, either containing the data of the neighboring block or values set by a boundary condition.

The simulations solve the three-dimensional Navier-Stokes equations for compressible flows together with an additional passive scalar concentration c. The state vector $\boldsymbol{W}$ (density ρ, energy E, velocities u, v, w in the three Cartesian directions x, y, z, respectively) is formulated in conservative variables, so that the following equations are solved for each cell i, j, k (index directions),

$$\begin{aligned}
\underset{(a)}{\frac{\partial \boldsymbol{W}_{i,j,k}}{\partial t}} &= \underset{(b)}{\boldsymbol{F}_{\text{conv},i,j,k}(\boldsymbol{W}_{i,j,k})} + \underset{(c)}{\boldsymbol{F}_{\text{visc},i,j,k}(\boldsymbol{W}_{i,j,k})} \\
&\quad - \underset{(d)}{\chi \times \left(\boldsymbol{W}_{i,j,k} - \boldsymbol{W}^{\star}_{i,j,k}\right)} - \underset{(e)}{\sigma_{i,j,k} \times \left(\boldsymbol{W}_{i,j,k} - \boldsymbol{W}_{\text{ref},i,j,k}\right)} , \\
\boldsymbol{W} &= \left[\rho,\, \rho u,\, \rho v,\, \rho w,\, \rho E,\, \rho c\right]^{\mathrm{T}} .
\end{aligned} \tag{1}$$

The individual terms (a)–(e) are described in the following.

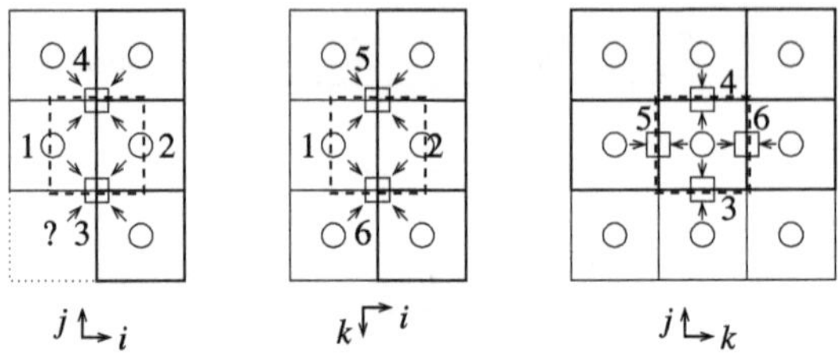

Fig. 3 Evaluation of surface center values of a staggered grid near block edges. Bold solid: interior cell; thin solid: ghost-layer cell; dashed: staggered cell; dotted: ghost-edge cell; circle: known value; square: interpolated value; question-mark: unknown value; number: surface number of staggered cell

2.2 Integration and Flux Evaluation

The time integration of Eq. 1 is done by a low-storage four-step Runge-Kutta scheme which is of second-order accuracy for non-linear equations [13]. The centered, fourth-order accurate approximation of the convective fluxes $\boldsymbol{F}_{\text{conv}}$ in skew-symmetrical form, cf. Eq. 1(b) [3], uses a five-point stencil that is reduced to a second-order accurate scheme with a three-point stencil at the boundaries.

In order to compute the second-order accurate discretization of the viscous fluxes $\boldsymbol{F}_{\text{visc}}$ Eq. 1(c), at the cell centers, the velocity gradients at the cell surfaces are required. (Three grids each staggered in one spatial direction are used to determine the gradients at all six surfaces of the cells.) They are evaluated using the velocities at the surface centers of a staggered grid that are equivalent to the cell centers of the original grid for two surfaces. The other four surface values are interpolated from the known values at the four surrounding cell centers of the original grid. Only at block edges or block corners, one or two values, respectively, are not known to the block since they reside neither inside the block nor in the ghost layers but in the ghost edges, cf. Fig. 3. In structured block-grid areas, these unknown ghost edge values can be copied from the diagonally opposing block for the cost of diagonal communication. However, this is not done since the communication overhead caused by a large number of communication events with a small amount of data would reduce the performance of the whole simulation. In addition, for unstructured zones, e.g. O-grids, the ghost edge value is ambiguous and therefore is extrapolated. Analogously, for the case of two inflow and outflow boundary conditions touching the same block edge, the unknown value is extrapolated. At the walls, the ghost edge is unambiguous and is obtained from the no-slip condition.

2.3 Turbulence Modeling

As stated in Sect. 2.1, LES are performed which require modeling of the small scales not represented on the grid. For this purpose, the Relaxation-Term model [11] of the Approximate Deconvolution Method [12] (ADM-RT) with a constant relaxation-coefficient χ in Eq. 1 is used. Previously, ADM has been implemented [16], parallelized [17] and now optimized for efficient computation. Referring to Eq. 1(d), the deconvolved filtered state vector $\boldsymbol{W}^\star$ is defined by

$$\boldsymbol{W}^\star := G_{3\mathrm{D}} * \underbrace{\sum_{\nu=0}^{N} (I - G_{3\mathrm{D}}*)^{\nu}}_{\text{deconvolution}} * \boldsymbol{W} = \sum_{\nu=0}^{N+1} \beta_{\mathrm{coeff},\nu} \times G_{3\mathrm{D}}^{\nu} * \boldsymbol{W} \, . \tag{2}$$

Commonly, a deconvolution order of $N = 5$ is used resulting in a six-fold application of the three-dimensional filter $G_{3\mathrm{D}}$ (identity operator I) [12]. The latter is implemented by applying the one-dimensional explicit filter G three times, once for each index direction $G_{3\mathrm{D}} = G_\mathrm{k} * G_\mathrm{j} * G_\mathrm{i}$. The filter is, e.g. for the first index direction i, defined as

$$G_\mathrm{i} * \boldsymbol{W}_{i,j,k} := \sum_{l=-\nu_\mathrm{l}}^{\nu_\mathrm{r}} \alpha_{\mathrm{coeff},i,j,k,l} \times \boldsymbol{W}_{i+l,j,k} \, . \tag{3}$$

A five-point stencil $\nu_\mathrm{r} + \nu_\mathrm{l} = 4$ is used which is asymmetric for boundary cells (e.g. $i = 0$: $\nu_\mathrm{l} = 1$) and symmetric with $\nu_\mathrm{r} = \nu_\mathrm{l} = 2$ for all interior cells.

2.4 Boundary Treatment

In the vicinity of the inflow and outflow boundaries, sponges [2] are employed, Eq. 1(e). They drive the solution towards the desired reference state $\boldsymbol{W}_\mathrm{ref}$. Non-zero σ-values select the regions and states where sponges are active, cf. Fig. 2. For sponges at the inflow, all but the energy, and for those at the outflow the pressure is modified.

As boundary conditions, no-slip, adiabatic conditions at walls, Dirichlet conditions prescribing all but the energy at inflow boundaries, and non-reflecting characteristic conditions [14] at outflow boundaries are chosen. The turbulent inflow data are generated by the Synthetic-Eddy Method (SEM) [8] using an extension to compressible flows [10]. These synthetic turbulence data, namely density and velocities, are fed in with a Dirichlet-type boundary condition.

2.5 Computational Environment and Measuring Procedure

All computations are carried out on the NEC SX-8 and SX-9 machines at HLRS. Information on some important system parameters are summarized in Table 1. To get the runtime and performance information of different code sequences, the Ftrace routines `ftrace_region_begin` and `ftrace_region_end` are called. Therefore, Ftrace is included during compiling and linking. The compila-

Table 1 Important system parameters of the NEC SX installations at HLRS

Series	# CPUs per Node	Peak performance per CPU P_{max} [GFLOPS]	Bandwidth [GByte/s] memory per CPU	Bandwidth [GByte/s] network per node
SX-8	8	16	64	8
SX-9	16	102.4	256	32

tion of the binaries is done on the front-ends for each architecture separately. The measurements presented compare results obtained on the two SX-architectures.

3 Performance Results and Flow Field Visualization

3.1 Sequential Performance of Individual Code Components

The simple geometry of case 1, a flat plate turbulent boundary layer (cf. Fig. 2) allows for a single-block topology. Therefore, parallelization overheads like load imbalances or communication time do not appear and the performance of the individual code components can be analyzed. In order to get statistical information of the flow, the field is sampled for averaging every time step. Besides simple averaging, a number of cross-correlations are computed too. The data depicted in Fig. 4 is averaged over 100 timesteps. The graph shows the relative 64-bit floating-point performance P/P_{max} (measured in FLOPS) as well as the time (in seconds) per timestep and cell the routine is working on. Since this block contains around 3.1 million grid cells, the performance rates obtained are rather high. For a computation with converged flow statistics, the grid should be split into multiple blocks to obtain a reasonable turn-around time of the simulation. In the following, the update of the ghost edges as described in Sect. 2.2 is not shown separately nor included in the boundary conditions since its share is below 1%.

For the SX-8, the relative performance of the routines acting on all cells cf. Eq. 1 is above 35%. This high value stems from the high fraction of vectorizable code and the long vectors that span through the whole field. Consequently, the boundary conditions (BC), acting only on planes of cells obtain a lesser performance of around 1%. The timings show that the computation of the sponge zones, cf. Eq. 1(e), and the statistics (stat) is at least one order of magnitude slower compared to the evaluation of Navier-Stokes (NS) equations, cf. Eq. 1(a)–(c), and of the turbulence model, cf. Eq. 1(d). Interestingly, the costs to solve the equations of fluid motion are only a bit higher than the costs to model the scales not represented on the grid. The boundary conditions require more than one order of magnitude more time per (boundary) cell than the evaluation of the Navier-Stokes equations. Fortunately, the number of cells being acted on is, for case 1, about 28 times lower than the total number of cells. Still, the total time spent on evaluating boundary conditions is remarkable, cf. Sect. 3.2.

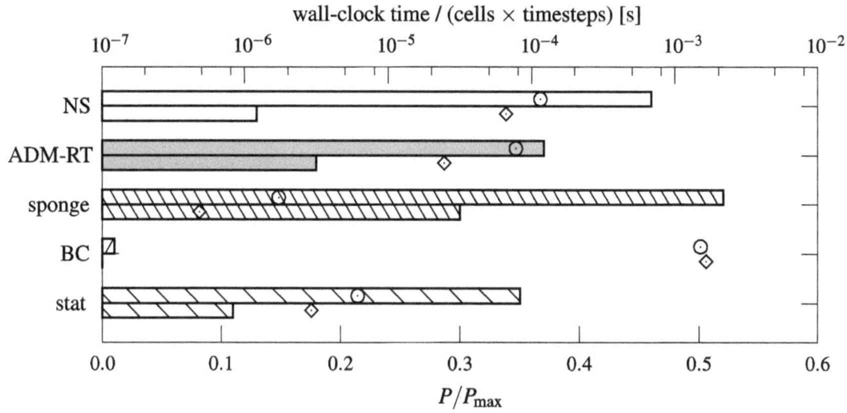

Fig. 4 Timing and performance of principal routines. Bars, upper abscissa: timing per cell; symbols, lower abscissa: relative performance $P/P_{\max}$; upper, circle: SX-8; lower, rhombus: SX-9

Considering case 1, the newer machine, SX-9, shows a lower level of relative performance for all routines. Therefore, the nominally offered more than six times higher performance cannot be used which is also observed comparing the timing. Nevertheless, there is a benefit in turn-around time for all routines except for the boundary conditions. Altogether, the computation on the SX-9 is just twice as fast as on the SX-8.

3.2 Overall Parallel Performance

Case 2 represents a film-cooling flow scenario, cf. Fig. 2. It consists of 90 blocks of different extent in physical space and index space. Most of the unstructured block topology, cf. Fig. 5, is given by the geometry. Nevertheless, the region above the plate additionally is split to optimize the load balancing for the use of 32 processors ($N_{\mathrm{p}} = 32$). The processors are evenly distributed across the minimum number of nodes. The simulations of 10 timesteps on an increasing number of processors using the same grid incorporating around 27 million cells are measured (strong scaling test).

The relative time share of principal routines averaged over the MPI-processes (one MPI process per processor) used is shown in Fig. 6. The computational loop is split into Navier-Stokes, turbulence model, boundary treatment (including boundary conditions and sponge zones) and communication part. The communication part includes all ghost-cell updates during repeated filtering (sixfold filtering in three index directions requires $3 \times 6 - 1 = 17$ ghost cell updates of the velocities between the filter applications) and time integration as well as the time that processors have to wait for each other due to load imbalances. For the SX-8, as found in the preceding section, solving the equations of fluid motion and turbulence modeling occupies most of the computing time. Already for the sequential case, a time share of around 5%

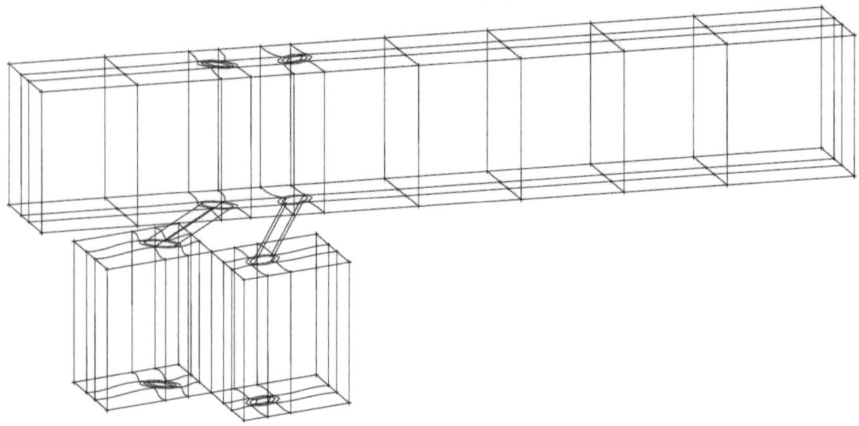

Fig. 5 Unstructured block-grid illustrating outer edges of the blocks

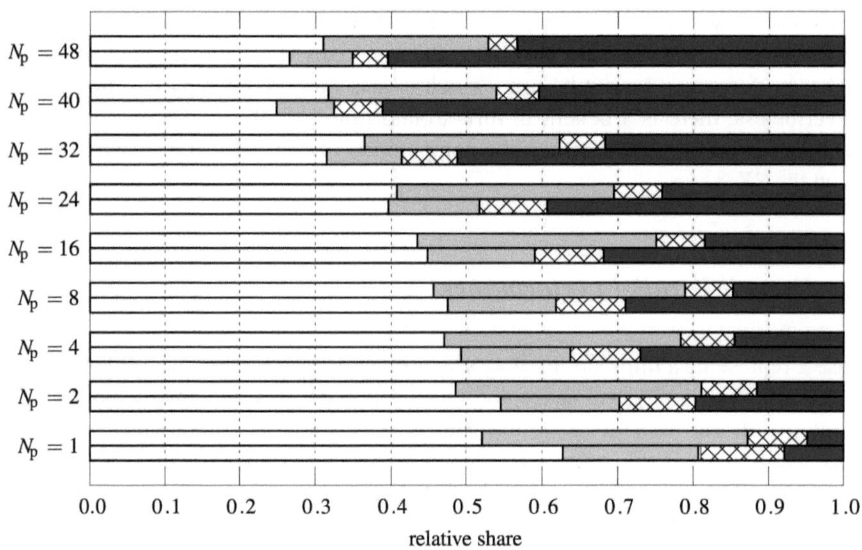

Fig. 6 Relative time share of principal routines. Upper: SX-8; lower: SX-9; empty: Navier-Stokes; lightly shaded: turbulence model; cross-hatched: boundary treatment (BC & sponge); darkly shaded: communication

for ghost-cell updates between the 90 blocks is observed. As discussed before, the low performance of the boundary condition computations plays a minor role. The evaluation of the Navier-Stokes equations, ADM-RT and the boundary treatment scales similarly. Increasing the number of processors the time share of communication increases up to over 40%. In this development, a jump is observed between 32 and 40 processors. This coincides with a kink in the speedup and will be discussed in the next paragraphs.

As apparent in Fig. 4, for the SX-9 the relative time share of the boundary treatment increases while the filtering performs relatively better compared to the SX-8. For all runs, the relative time spent on communication is higher since the data exchange does not depend on the processor but on the network and the memory bandwidth, that both are not six times faster. The configuration using 40 processors distributed across three nodes performs worse, in terms of communication, than using 48 processors. This might be caused by the unequal distribution of 13, 13 and 14 used processors on three nodes.

The speedup S is the ratio of the time t the parallel computation ($N_p > 1$) takes compared to the sequential one ($N_p = 1$),

$$S(N_p) := \frac{t(N_p = 1)}{t(N_p)} . \tag{4}$$

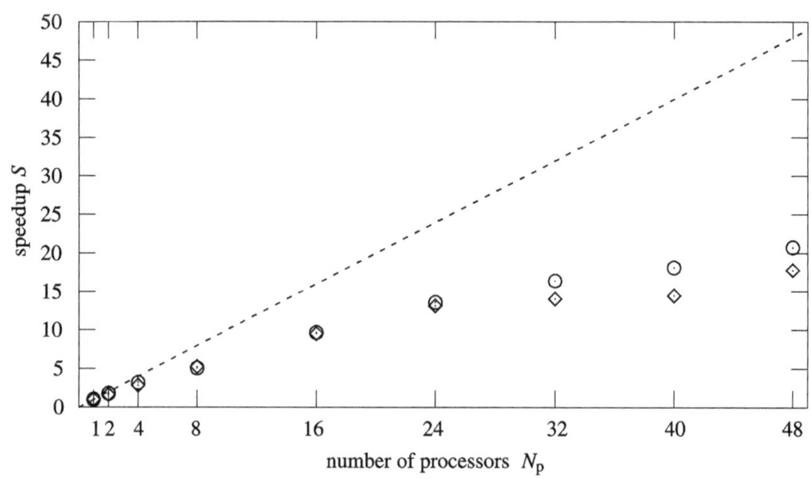

Fig. 7 Speedup of case 2. Symbols: measurement; dashed: ideal speedup; circle SX-8; rhombus: SX-9

Consequently, the ideal speedup would be $S = N_p$. For case 2, Fig. 7 illustrates the efficiency of the parallel runs based on wall-clock data of 10 timesteps. For the SX-8 a nearly linear speedup with a sub-ideal slope up to 32 processors appears. For the latter configuration, most MPI-processes contain exactly one of the big blocks above the plate, cf. Fig. 5, and the others share the remaining blocks that are somehow evenly distributed among them. Beyond 32 processors, the smaller blocks do not fill up the remaining processors sufficiently well, so that they have to idle until the computation of the bigger blocks is finished. The SX-9 shows an almost identical behavior up to 24 processors. Beyond that limit, the speedup is somewhat lower compared to the SX-8. For 40 processors, a speedup is hardly detectable. To obtain a higher speedup beyond 32 processors the grid needs to be split-up into more blocks improving the load balancing but introducing additional communication overhead.

3.3 Flow Field Visualization

To get an impression of the mean flow field, Fig. 8 illustrates streamlines. A number of interesting facts can be educed from this figure. First, the fluid jets ejected through the upstream and downstream holes do not mix intensely, since only few streamlines of the upstream hole are present in the region pointed at by the downstream hole and vice versa. A low level of mixing occurs in the vicinity of the downstream hole. Second, far downstream the upstream jet covers a narrow spanwise and a high wall-normal region whereas the downstream jet covers a broader spanwise and a flatter wall-normal region. The upstream jet faces an undisturbed boundary layer and due to its compound-angle injection one dominant clockwise vortex establishes. In contrast, the downstream jet is ejected next to this strong vortex and pushed towards the wall by the downwash induced there. Furthermore, the temperature contours show a significant coverage of the plate by the coolant. The temperature at the centerplane shows that the region of the first coolant jet detachment is small and that the second jet does not detach at all. The different cross-stream sections show a downstream broadening of the film.

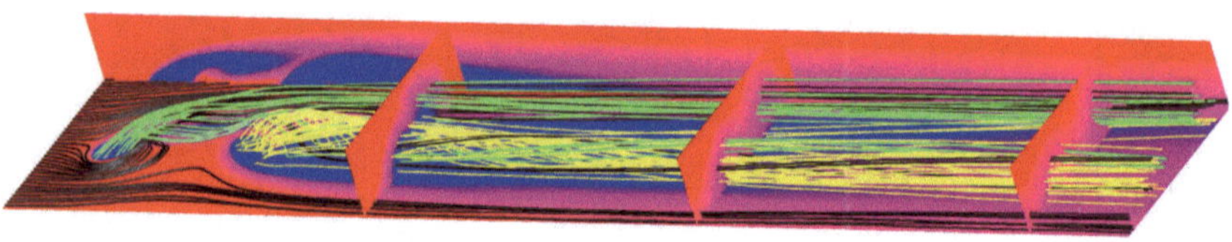

Fig. 8 Mean flow field visualized by streamlines. Fluid ejected through the upstream hole (green), through the downstream hole (yellow) and crossflow (black). Contours: temperature (high amount of cooling: blue; low: red) at the wall, at cross-stream planes and at center plane (laterally shifted)

4 Conclusions

Computational aspects of boundary-layer and film-cooling simulations with LES using the code NSMB have been discussed. First, the performance of principal routines has been analyzed for a sequential simulation of a canonical flat plate turbulent boundary layer (case 1). Then, a strong scaling test of the film cooling setup (case 2) has been conducted. Also, the shift of the share of the most important routines for an increasing number of processors has been documented. The measurements have been compared for the two architectures NEC SX-8 and SX-9.

Concerning the SX-8, from the boundary-layer simulation using a single-block grid (case 1) it is concluded that the principal routines solving the equations of fluid motion and modeling the turbulence have a high relative performance above 35% of the theoretical peak. The performance of the boundary condition computations is as low as 1%, however, their share of the total computation is small.

For the SX-8, the parallel film-cooling simulations presented (case 2) show that the share of the communication increases with the number of processors to more than 40% for 48 processors, while the other routines share thc remaining computational time with a constant ratio for different numbers of processors. Up to 32 processors, more than half of the ideal speedup is obtained. Obviously, the loss is mainly caused by the time spent on communication.

The SX-9, nominally providing a more than six times higher performance, performs only twice as fast as the SX-8 for the considered case. The relatively slow communication causes its relative share to increase to over 50% already for 32 processors. In the end, the unequal increase of computational performance compared to memory access and network access is not beneficial for tightly coupled parallel applications performing frequent communication, as for the considered case 2.

Considering the increase in core numbers of today's supercomputing facilities, the trend observed from SX-8 to SX-9 is expected to continue. When using the ADM-RT model with a six-fold filtering of the whole field, spending most of the time waiting for data to be communicated is inevitable. To avoid this, it could be considered to reduce the order of deconvolution.

Acknowledgements This work is partly funded by the Swiss National Science Foundation (SNF) with the project number 200020-116310. All computations are conducted at the High Performance Computing Center Stuttgart, Germany (HLRS) with resources granted by the DEISA Consortium, co-funded through the EU FP6 project RI-031513 and the FP7 project RI-222919, within the DEISA Extreme Computing Initiative under the project acronym FCool2. We gratefully acknowledge the support by S. Haberhauer of NEC Germany.

References

1. Ahn, J., Jung, I.S., Lee, J.S.: Film Cooling from Two Rows of Holes with Opposite Orientation Angles: Injectant Behavior and Adiabatic Film Cooling Effectiveness. Int. J. Heat Fluid Flow 24(1), 91–99 (2003). DOI: 10.1016/S0142-727X(02)00200-X
2. Bodony, D.J.: Analysis of Sponge Zones for Computational Fluid Mechanics. J. Comput. Phys. 212(2), 681–702 (2006). DOI: 10.1016/j.jcp.2005.07.014
3. Ducros, F., Laporte, F., Soulères, T., Guinot, V., Moinat, P., Caruelle, B.: High-Order Fluxes for Conservative Skew-Symmetric-like Schemes in Structured Meshes: Application to Compressible Flows. J. Comput. Phys. 161(1), 114–139 (2000). DOI: 10.1006/jcph.2000.6492. http://www.sciencedirect.com/science/article/B6WHY-45FC8JY-4S/2/d96be1945d6770e7b755665fc737f30f
4. Goldstein, R.J.: Film Cooling. In: Irvine, T.F. Jr., Hartnett, J.P. (eds.) Advances in Heat Transfer vol. 7, pp. 321–379. Elsevier (1971). DOI: 10.1016/S0065-2717(08)70020-0. http://www.sciencedirect.com/science/article/B7RNJ-4S818YJ-B/2/8c14c944f842c9e90ee56af358b12c72
5. Gräf, L., Kleiser, L.: Large-Eddy Simulation of Double-Row Compound-Angle Film-Cooling: Setup and Validation. Comput. Fluids (2010). Accepted
6. Han, J.C., Dutta, S., Ekkad, S.: Gas Turbine Heat Transfer and Cooling Technology. Taylor & Francis (2001)
7. Haven, B.A., Yamagata, D.K., Kurosaka, M., Yamawaki, S., Maya, T.: Anti-Kidney Pair of Vortices in Shaped Holes and Their Influence on Film Cooling Effectiveness. Technical Report 97-GT-45, ASME (1997)

8. Jarrin, N., Benhamadouche, S., Laurence, D., Prosser, R.: A Synthetic-Eddy-Method for Generating Inflow Conditions for Large-Eddy Simulations. Int. J. Heat Fluid Flow 27(4), 585–593 (2006). DOI: 10.1016/j.ijheatfluidflow.2006.02.006
9. Lee, S.W., Kim, Y.B., Lee, J.S.: Flow Characteristics and Aerodynamic Losses of Film Cooling Jets with Compound Angle Orientations. ASME J. Turbomach. 119, 310–319 (1997). DOI: 10.1115/1.2841114
10. Pritz, B., Magagnato, F., Gabi, M.: Inlet Condition for Large-eddy Simulation Applied to a Combustion Chamber. In: Conf. Modelling Fluid Flow, Budapest, Hungary 2006
11. Schlatter, P.: Large-Eddy Simulation of Transition and Turbulence in Wall-bounded Shear Flow. PhD thesis, ETH Zurich (2005). Diss. ETH No. 16000
12. Stolz, S., Adams, N., Kleiser, L.: The Approximate Deconvolution Model for Large-Eddy Simulations of Compressible Flows and its Application to Shock-Turbulent-Boundary-Layer Interaction. Phys. Fluids 13(10), 2985–3001 (2001). DOI: 10.1063/1.1397277
13. Swanson, R.C., Turkel, E.: Multistage Schemes With Multigrid for Euler and Navier-Stokes Equations. Tech. Paper 3631, NASA Langley (1997)
14. Thompson, K.W.: Time Dependent Boundary Conditions for Hyperbolic Systems. J. Comput. Phys. 68, 1–24 (1987). DOI: 10.1016/0021-9991(87)90041-6
15. Vos, J.B., van Kemenade, V., Ytterstrom, A., Rizzi, A.W.: Parallel NSMB: An Industrialized Aerospace Code for Complete Aircraft Simulations. Parallel Comput. Fluid Dyn.: Algorithms and Results Using Advanced Computers, 49 (1996)
16. von Kaenel, R.: Large-Eddy Simulation of Compressible Flows Using the Finite-volume Method. PhD thesis, ETH Zurich (2003). Diss. ETH No. 15255
17. Ziefle, J.: Large-Eddy Simulation of Complex Massively-separated Turbulent Flows. PhD thesis, ETH Zurich (2008). Diss. ETH No. 17846

Large Eddy Simulation of Wind Turbulence for Appropriate Urban Environment

Tetsuro Tamura

Abstract This study shows the numerical model which has the actual shape of the urban surface in order to analyze the wind turbulence in a city by large eddy simulation (LES) technique. Predictive accuracy of LES for the urban heat island, the hazardous gas dispersion or failure of buildings by wind gust was examined. Also, the numerical results were provided for realizing appropriate urban environment.

A numerical model of an urban city is devised to study different types of scenarios. The actual shape of the urban city is represented using the geographic information science (GIS) data. As an environmental problem in the atmospheric flows, the urban heat island or the hazardous gas dispersion is considered. Meanwhile as a wind disaster problem, human impact on high wind or failure of buildings and structures by wind gust is examined for the mitigation against typhoon attack to a city. LES is capable of providing accurate time-sequential physical quantities and estimating the peak values of various impacts. Based on a plenty of data for wind turbulence obtained by LES, the appropriate method for sustaining comfortable urban environment is discussed. Since this numerical approach generally requires a large amount of computational time, cost and capability, we have to formulate a requirement for the size of computational domain and resolution, and the time scale for evaluation of statistics.

For constructing an appropriate LES model of atmospheric turbulent flow, the technological potential should be examined concerning complex boundary layer over various underlying surfaces. Nowadays, we can employ a computational model which directly reproduces actual and complex configurations at surface of city, such as buildings, structures, vegetation and trees on various kinds of terrain. This study focuses on the interior of the urban canopy where near-ground flows are very complex and unsteady due to flow separation and vortex shedding induced by building roughness elements.

Tetsuro Tamura
Tokyo Institute of Technology, 2-12-1 Ookayama, Meguro-ku, Tokyo 152-8550 Japan
e-mail: tamura.t.ab@m.titech.ac.jp

M. Resch et al. (eds.), *High Performance Computing on Vector Systems 2010*,
DOI 10.1007/978-3-642-11851-7_15, © Springer-Verlag Berlin Heidelberg 2010

A detailed comparison of LES results with observational data should be carried out for the model validation. Generally, LES cannot predict the absolute wind speed on specified date at a specified site, so it provides only a value relative to the reference wind speed. In order to introduce absolute values in LES, the meso-scale meteorological model is utilized. Namely, combining meteorological and LES results at a reference location, we can generate the approaching turbulent flow with specific properties over actual urban area. As a meteorological event, this study picks up high wind during cyclogenesis or typhoon and low wind by local circulation such as a sea breeze. Wind fields on urban scale are simulated by the meteorological model, while turbulent flow fields inside the urban canopy are computed by LES that explicitly incorporates effects of actual building shapes. To realize a very large computation using fine grid resolution by the combined model approach, details of the method for different practical applications have been provided.

Required computer resources for practical solutions by this model are examined in order to clarify the trends in high-performance computing for wind-related problems in urban areas. The combined analysis shows results consistent with observational data, and also exhibits details of local gust wind and sudden changes of canopy winds. This study suggests that the combined method could be a powerful tool for mitigating a severe wind disaster and a degradation of atmospheric environment in urban areas.

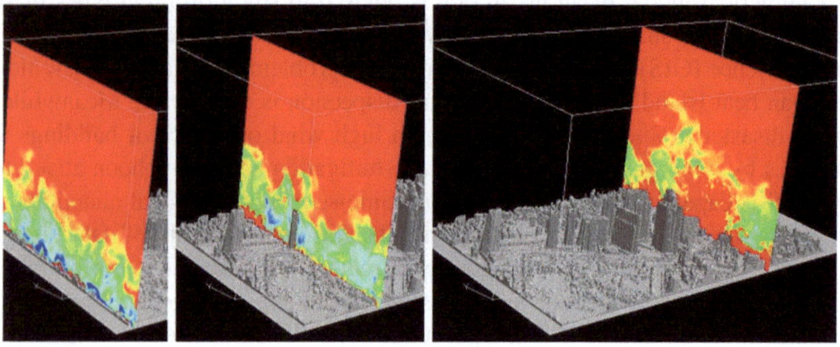

Fig. 1 Instantaneous temperature fields in urban area

MIX
Papier aus verantwortungsvollen Quellen
Paper from responsible sources
FSC® C105338

If you have any concerns about our products,
you can contact us on
ProductSafety@springernature.com

In case Publisher is established outside the EU,
the EU authorized representative is:
Springer Nature Customer Service Center GmbH
Europaplatz 3, 69115 Heidelberg, Germany

Printed by Libri Plureos GmbH
in Hamburg, Germany